PEACOCK BOOKS
Editor: Kaye Webb

Finding Fossils

Life has existed on Earth for well over 3,000 million years, and during that vast span the remains of countless animals and plants have become trapped and preserved in rock. As a result, we are left with a wonderful record of life on Earth, including many species that are now extinct.

In Britain alone millions of fossils are embedded in the rocks. So anyone possessing some basic information about palaeontology, as the study of fossils is called, ought to be able to build up an interesting collection. But most books for the inexperienced collector do not give precise enough instructions on where to look for specimens; nor do they show specimens in a realistic condition.

Roger Hamilton and Allan N. Insole decided to do something about this. During one summer they visited five areas of Britain where you might expect to find fossils: the Kent coast, the Dorset coast, the Isle of Wight, the Bristol and Gloucester region, and the Yorkshire coast. They took a photographer along with them, and kept a detailed record of the places they visited and the fossils they found. The result is a five-day fossil-collecting diary, in word and picture, which directs the reader with precise instructions to about thirty good sites in these areas and shows the results of the authors' collecting there.

This diary forms the backbone to the book, but there are also introductory sections explaining the different kinds of fossils and their value, as well as a concluding chapter on how to treat and classify the specimens you find. With its realistic photographs, all specially taken for this book, and its many diagrams, this is an ideal introduction to fossil collecting.

Roger Hamilton and Allan N. Insole both hold doctorates from the geology department of Bristol University, and both now work in museums, Dr Hamilton at the British Museum (Natural History) in London and Dr Insole on the Isle of Wight.

Finding Fossils

Roger Hamilton and
Allan N. Insole

Penguin Books

Penguin Books Ltd,
Harmondsworth, Middlesex, England
Penguin Books, 625 Madison Avenue, New York, New York 10022,
U.S.A.
Penguin Books Australia Ltd,
Ringwood, Victoria, Australia
Penguin Books Canada Ltd,
2801 John Street, Markham, Ontario, Canada L3R 1B4
Penguin Books (N.Z.) Ltd,
182–190 Wairau Road, Auckland 10, New Zealand

First published 1977
Published simultaneously in hardback by Kestrel Books

The diagrams were drawn by Michael Strand from information
provided by the authors. The photographs were taken by Imitor
(Colin Keates and Peter Green)

Filmset in Monophoto Ehrhardt by
Richard Clay (The Chaucer Press) Ltd, Bungay, Suffolk,
and printed in Great Britain by
Fletcher & Son Ltd, Norwich

Contents

Introduction

This is a book about fossil collecting. It is a practical book and its use will help the collector to gather a large collection of fossils in which plants and most of the main groups of animals are represented. Brief details of the fossil groups are given and important features used in their identification are described and explained. We have kept the use of technical terms to a minimum but where necessary these are explained and used so that the reader will be able to refer more easily to detailed books of geology and palaeontology.

Fossils are common and can be easily collected in many parts of Britain, but it is important to know where to look and to be able to recognize them in the rock. To help readers with this we have visited and collected fossils at a number of sites throughout the country. Our account of the collecting, the sites, the problems encountered and the fossils that we found forms the main part of this book. Photographs used in the book were all taken during our visits or show fossils that we collected. Collecting was done mainly by the two authors and by the photographers, Colin Keates and Peter Green; Melvin Green helped us on days 1, 2 and 4. We allowed ourselves five days for collecting; these were spaced out over the summer of 1975. During this time we visited twenty-eight sites and collected over 500 specimens.

By limiting ourselves to five days our collecting time was of course restricted, but in each case representative collections were made. Most of the sites were along cliffs and beaches, as these are the places most likely to be reached by the amateur collector. Tidal conditions were not always perfect. For example, it is impossible to visit three sites in north Kent or in Yorkshire during the same day and find the tide just right in each place. The weather was sometimes bad, but the collector who waits only for good weather may never make a good collection; and anyway fossil collecting is a good way to spend the time when it is too cold for swimming or sunbathing. The reader will probably spend longer on each site than we did and will therefore be able to get a better collection than ours. So even if only a few sites are visited a good basic collection can be built up. This experience should quickly enable the collector to discover sites for himself or to work from more detailed maps and site descriptions.

1   First Steps

Before you go out to collect there is some essential information of which you need to be aware. An understanding of the kinds of rocks and the groupings of animals and plants will help you to locate fossils and to make preliminary identifications; and details of the fossils and technical terms used in their description will help you to use other collecting guides.

WHAT IS GEOLOGY?

Geology is the science of the Earth. It deals with the origin, structure, composition and history of our planet and the processes that have affected it and are affecting it today. In this book we are concerned mainly with that branch of geology that deals with fossils and is called *palaeontology*. However, to find fossils and understand how they are formed we must also look briefly at some of the rocks that form the earth.

TYPES OF ROCKS

The rocks of the earth consist of eight main elements – aluminium, calcium, iron, magnesium, oxygen, potassium, silicon and sodium. These elements are combined in various ways to form definite chemical compounds that are called *minerals*. Masses of single minerals or mixtures of different minerals form rocks. Very many different rocks occur on the surface of the earth but only certain varieties contain fossils. It is therefore useful if the collector can recognize the unproductive types, so that time is not wasted searching in the wrong spots. Three broad classes of rocks are identified according to the way in which they are formed. These are called igneous, sedimentary and metamorphic rocks.

Igneous rocks

Igneous rocks are produced when molten rock called magma solidifies or crystallizes in the crust of the earth. Magmas originate at depths of 50 to 250 km below the surface of the earth, and they rise through fractures and other weak points in the Earth's crust. These magmas may cool and crystallize in the crust and produce *intrusive* rocks, or they may reach the surface and erupt as lavas through volcanoes to form *extrusive* rocks. Although igneous rocks vary in colour and texture they can usually be identified because they consist of a mass of interlocking crystals that are visible with a hand lens or with the naked eye. Also many of them do not show any obvious banding: ' banding' is the term used when rocks of different colours form bands or lines – these are often wavy and they vary in thickness.

Sedimentary rocks

Sedimentary rocks are formed when rock fragments and minerals are deposited at normal temperature on land or under water. One of the most obvious characteristics of sedimentary rocks is their distinctive layering, which is called stratification or bedding. *Bedding planes* are divisions separating individual layers. The texture and minerals composing the rock are used to divide sedimentary rocks into two groups – clastic and chemical.

Clastic rocks are those built up from fragments of other

Sedimentary rocks showing bedding planes at Charmouth, Dorset

rocks, and clastic sediments are those that show evidence of mechanical transport and deposition of rock fragments by wind or water currents. These rock fragments and minerals result from the weathering of other rocks and the fragments may be worn during transport so that the grains are rounded. During the formation of a deposit the grains become sorted according to size and weight, with the large and heavy particles being deposited first. So clastic sediments are divided, according to grain size, into gravels, sands, silts and clays. When these sediments become compacted by pressure or cemented by chemicals they form rocks that are called conglomerates, sandstones, siltstones and claystones.

Chemical rocks are those consisting either of minerals deposited from solution by chemical reactions or of plant remains or the skeletons of animals. They are subdivided according to their composition. The most familiar chemical sedimentary rocks are limestones (which include chalk), iron-stones, iron pyrites, flint, rock salt, gypsum, coal and peat.

Almost all fossils are found in sedimentary rocks, which are therefore the most important group of rocks for the collector, though not all sedimentary rocks yield fossils.

Metamorphic rocks

If a rock is heated or if it is subjected to increased pressure (or if both of these things happen) deep in the Earth's crust, its chemical composition and minerals may be altered. This change is called metamorphism. The nature of the change varies, but it may involve the recrystallization of the rock: the breakdown of the original crystals of rock and the formation of new crystals. These may be small or sometimes much larger than the original crystals. Alternatively it may involve the formation of new minerals and the development of wavy structures that result from minerals, especially mica, changing position so that they are arranged in parallel lines. The type of rock that results from metamorphism depends on the composition of the original rock and the conditions under which the changes occurred. For example, a limestone that is subjected to heat and pressure may change into marble, while a claystone will change to a slate or schist. Most metamorphic rocks have a crystalline texture and there are often conspicuous flakes of mica and irregular bands.

Most metamorphic rocks do not contain fossils, but occasionally specimens can be collected from slates. These fossils are usually distorted and may be unrecognizable as a result of the stresses to which the rock has been subjected.

TYPES OF FOSSILS AND HOW THEY FORM

Fossils are any remains or evidence of past life that are

preserved in the rocks. There are in general two kinds of fossils: those representing the actual animal or plant, called 'body fossils', and those that represent structures which were produced by the activity of the animals or plants in the sediments, called 'trace fossils'.

When an animal or plant dies, its remains are either eaten by scavengers or they rot away, so that after only a relatively short time even the hard parts such as shells or bones will have vanished. Of course, the soft parts decay much more quickly than the hard parts. For an organism to be fossilized, parts of it must be preserved. This usually occurs as a result of burial shortly after death, since burial protects the remains from scavengers and may also prevent bacterial action.

The sediments that cover the remains must be suitable for their preservation. They must first form rocks which will preserve the remains as they undergo the processes of fossilization described below. Then the rocks must themselves survive and of course be exposed to allow the discovery of the fossils they contain.

Sediments are most usually deposited under water, either in the sea or in rivers and lakes. As a result most fossils represent water-living animals or plants, while most fossils of land animals are found in rocks that were deposited under water. The presence of hard parts also greatly increases the chances that an animal will be fossilized. As a result of these processes we tend to get an unbalanced idea of the life of the past from our studies of fossils. For example, molluscs and brachiopods (see p. 40) are very well represented in most rocks that were formed in the sea, whereas jellyfishes and worms are rare. There is, however, no reason to think that these soft-bodied animals were any less common in the past than they are today. Their poor representation as fossils simply reflects the fact that their bodies usually lack hard parts.

Fossils may consist only of the original material of the animal or plant. Many fossils younger than one million years are like this, and the survival of fossils without marked alteration occurs most frequently in the preservation of shells, bones and teeth which are already very hard and resistant to break-down by other natural processes. In some cases minerals are removed from the remains during fossilization. This most usually occurs when salts in the shell or bone dissolve in water as it filters through the remains and through the surrounding sediments, a process known as *leaching*. The shells from the Thanet Beds at Herne Bay (see p. 78) have been leached so that they are lighter and more brittle than modern shells.

Usually, however, the original material of the animal or plant becomes impregnated by minerals which enter in solution from the surrounding sediments. This process is called *mineralization* and it often makes the fossils hard or stony. Silica, calcite or pyrites are often involved in mineralization. The bones from the Aust bone beds (see p. 128) have been heavily mineralized, so that they are harder and heavier than modern bone.

When the minerals have been added gradually and the original material of the remains is also preserved, microscopic details of structure can be studied in the fossils. Sometimes the original material of the remains has been replaced entirely by minerals. This may have happened very slowly, even molecule by molecule, so that microscopic details of structure have been preserved, though in a different material. This process is known as *replacement*. The silicified shells from the Oldhaven Beds of Bishopstone Glen (see p. 79) are fossils resulting from this process. Replacement may also occur very quickly, in which case fine detail of structure is lost.

If the hard parts are dissolved by percolating water after burial, a hollow may be left in the rock so that only an impression of the original material remains. This may represent either the internal or external appearance of the remains and such impressions are called *moulds*. Moulds may later become filled with minerals that enter from solution and form a *cast* of the original remains. Casts are usually preserved in calcite or silica.

Some fossils result from the breakdown of the original

material until only a carbon film remains. This process is called *carbonization* and occurs, for example, in the fossilization of many plants, particularly in coals. Carbonized fossils are black and shiny and are usually flattened along the bedding planes of the enclosing rocks. Thus the original three-dimensional shape of the organism is lost.

Moulds and casts are really only traces of past life, and as such they are sometimes grouped as 'trace fossils'. However, the term 'trace fossil' more usually applies to evidence of past life such as borings (p. 36), burrows (p. 51), tracks, footprints and coprolites, which are fossilized dung.

Most fossils consist only of the hard parts of animals or plants, but occasionally and under very special conditions delicate or soft parts may be preserved, and in a few cases complete animals are fossilized. For example, delicate insect

Burrows on a fallen block at Eype, Dorset

wings may be found in the fine-grained rocks of the Bembridge Marls (Oligocene) at Thorness Bay on the Isle of Wight, while impressions of jellyfishes are known from similarly fine-grained rocks in other parts of the world. Complete insects may be preserved in amber, which is itself fossilized resin, while animals such as woolly rhinoceros and mammoths have been found pickled in oil and brine or frozen in the permafrost of Siberia and North America.

THE USES OF FOSSILS

Fossils can be used to find the relative ages of sedimentary rocks, and they are also used to give evidence about conditions in the past and about the history of animals and plants.

Modern animals and plants are adapted to live in particular conditions; this was also true of animals and plants that are now known only from their fossil remains. For example, sea urchins occur today only in marine conditions and they cannot tolerate fresh or even brackish water. There is no reason to suppose that sea urchins of the past were any different in this respect, so if we find rocks containing fossil sea urchins we can be sure that they were formed in the sea. By piecing together small pieces of information such as this it is possible to study complete environments. This sort of study is called *palaeoecology* and is one of the most exciting recent developments in palaeontology. Careful analysis of fossil communities can yield information not only about the general environment but also about factors such as the speed of water currents, depth and climates. When information such as this is combined from wide areas, even the geography of the past can be reconstructed.

The study of fossils also shows the way in which the life of the past has changed with time. There are many gaps in the fossil record, but the evolution of many animal and plant groups can be understood with the help of fossils.

GEOLOGICAL TIME

Geologists and astronomers now believe that the earth was formed about 4,500 million years ago, although the oldest rocks so far identified are about 3,500 million years old. It is important for geologists to know the relative ages of the rocks with which they work, and this can be done by the application of a few simple rules. Sometimes it is even possible to give the age of a single rock quite accurately. The two main principles that are used were suggested by an English surveyor and civil engineer called William Smith (1769–1839), who is rightly known as the father of English geology.

The first rule states that a layer of sedimentary rock must be younger than the layer upon which it rests. This is always true if it can be shown that the beds have not been turned upside down by earth movements. By tracing beds from one locality to another it is theoretically possible to build up a sequence of rock layers or 'strata', even though only a small part of that sequence may be visible at any one place. There are certain difficulties in applying this law, since beds do in fact change their character from place to place and it may therefore be impossible to trace or even identify a bed or group of beds over any great distance. Geologists call groups of beds with similar characteristics *formations*. The Chalk and the Gault Clay are well-known formations.

The second rule states that each layer, or group of layers, of sedimentary rock contains a characteristic collection of fossils that will differ from that occurring in layers above or below. Thus if we find that beds in two different localities contain the same assemblage of fossils we can suggest that they are of approximately the same age. This use of fossils to compare the ages of rock sequences is called *correlation*. It allows the relative ages of rocks to be worked out, but it can never give the actual age of a rock in years. In practice some fossils are not used in correlation because they are either rare, have limited geographical distribution or have very long time ranges.

By using the above two rules carefully geologists have been able to arrange rocks and formations according to their relative ages and to recognize rocks of similar ages in different parts of the world. As rocks deposited in the same time period differ from place to place, the early geologists found it necessary to use units of time that were not related to the rock types in their investigations. The result was the stratigraphical column shown opposite, which is now used throughout the world. The names of the units of this time scale, the periods and the eras were coined somewhat haphazardly by nineteenth-century geologists and they therefore have a variety of derivations.

It is important to realize that although periods and eras are recognized all over the world, geological formations have only limited geographical ranges. The number and character of the formations from a particular period vary from place to place even in a relatively small area like the British Isles. For example, rocks of the Jurassic period which we collected in Dorset (p. 89) and Yorkshire (p. 143) are very different in character, thickness and fossil content, although both sequences were laid down during approximately the same period of geological time.

The methods described above provided a relative time scale that geologists could use, but they did not give an actual age in years. However, with the discovery of radioactivity a technique became available by which rocks could be dated in years. This method depends on the fact that a radioactive element changes or 'decays' at a rate which is constant and is not influenced by the conditions of the surroundings. As they are formed, certain minerals 'trap' small amounts of radioactive elements, which then decay. Provided that the rate of decay is known and that the amount of radioactive element remaining in the mineral can be found, then by comparing this with the products of the decay it is possible to work out the length of time that has elapsed since the rock was formed. This technique is known as

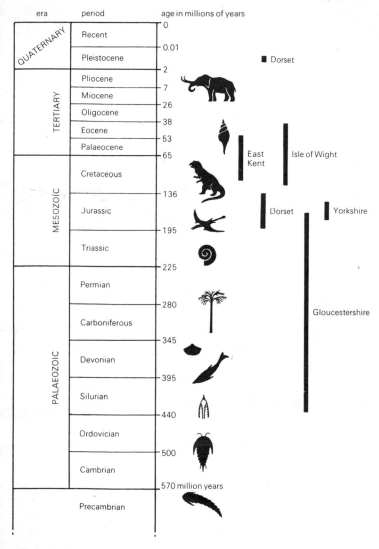

era	period	age in millions of years
QUATERNARY	Recent	0
		0.01
	Pleistocene	
		2
TERTIARY	Pliocene	
		7
	Miocene	
		26
	Oligocene	
		38
	Eocene	
		53
	Palaeocene	
		65
MESOZOIC	Cretaceous	
		136
	Jurassic	
		195
	Triassic	
		225
PALAEOZOIC	Permian	
		280
	Carboniferous	
		345
	Devonian	
		395
	Silurian	
		440
	Ordovician	
		500
	Cambrian	
		570 million years
	Precambrian	

Dorset

East Kent   Isle of Wight

Dorset   Yorkshire

Gloucestershire

Geological time-scale. The diagram shows the relative ages of rocks at the sites where we collected radiometric dating. Unfortunately it can only be applied to certain rocks, mainly of the igneous and metamorphic types. However, the age of sedimentary rocks can be discovered if they lie between beds of lavas or volcanic ashes, as these can be dated, so giving a maximum and minimum age for the sedimentary rocks in between.

GEOLOGICAL MAPS

A geological map shows the distribution and ages of the different kinds of rocks that lie under the soil. In the United Kingdom geological maps are published by the Institute of Geological Sciences and are coloured overlays of normal Ordnance Survey maps. In addition to these 'official maps', other maps giving more detailed information can often be found in geological guide books (see p. 158).

It is important to understand that geological maps do not indicate that the rocks are exposed at the surface in a particular area. In reality the bed rock will be visible at only a very few points in an area; elsewhere it will be covered by soil and vegetation. For this reason geological maps are not very helpful for initially finding possible collecting sites, but they can be used to find the stratigraphical position of a known site. This may be especially useful in cases where there are temporary exposures such as trenches, building sites or during road construction.

GEOLOGICAL STRUCTURES

The rocks that outcrop in the British Isles are very varied in their ages, ranging from at least 2,600 million years old in parts of the Outer Hebrides to the most recent ones, which have been laid down during the last few thousand years in some river valleys or estuaries. (A geological map of the British Isles

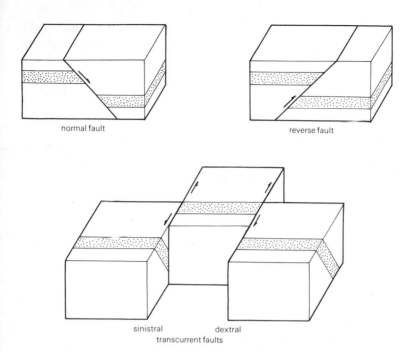

normal fault

reverse fault

sinistral   dextral
transcurrent faults

Diagrams of stratified rocks displaced by the three main types of faults

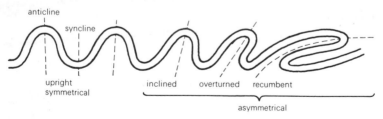

anticline

syncline

upright
symmetrical   inclined   overturned   recumbent

asymmetrical

Shapes of folds in profile

is shown on page 65.) Since many of these rocks were laid down as horizontal beds it may seem strange that they are not still piled up like the leaves of a book, with the oldest at the bottom and the youngest at the top. However, in geological terms the Earth is not a quiet place and rocks may be disturbed by faulting or folding. When faulting occurs, rocks are fractured and the blocks on either side of the fracture move in relation to each other. During folding, the layers of rock are bent and distorted in varying degrees. Areas that have been faulted or folded may then be eroded, and this will expose beds of rock that were previously buried deep below the surface. As a result, in most of the cliffs and quarries where fossils are to be found the rocks are no longer in horizontal beds but slope or 'dip'. This enables us to examine a succession of rocks rather than a single layer, and we can therefore obtain fossils representing a selection of geological time.

SCIENTIFIC NAMES AND CLASSIFICATION

Most familiar animals and plants have popular names. These may be accurate and define a single kind of organism throughout the world (for example, 'horse'), or they may be vague – 'robin', for example, refers to quite different birds in Britain and North America. Of course popular names change in different languages, and therefore no popular name is used throughout the world. Also many rarer animals and plants do not even have popular names, while others may cover many species. Most fossils do not have popular names.

To overcome these problems animals and plants have been given scientific names. These are internationally used, clearly defined and accurate. They are either in Latin or they are Latinized versions of names from other languages. When a scientist discovers a new animal or plant he names it, so all known species of animals and plants have scientific names.

A species is a group of animals that are interfertile; that is, they are capable of breeding together and producing fertile offspring. The scientific name of a species consists of two Latinized words. For example:

man	–	*Homo sapiens*
wolf	–	*Canis lupus*
pig	–	*Sus scrofa*

The second of the two names is called the trivial name and can only be used with the first or 'generic' name. The generic name refers to the genus (plural 'genera'), which is a group of similar species. The genus *Equus* – horses – includes the domestic horse (*Equus caballus*), the zebra (*Equus zebrinus*) and the donkey (*Equus asinus*). Note that the scientific name is printed in italics; and it should always be underlined in handwriting.

Genera are grouped into families, and these are in turn grouped into orders. For example, the family Felidae – cats – includes all species of cats in the genus *Felis* as well as the lions and tigers, which are in the genus *Panthera*, and the cheetah, which is in the genus *Acinonyx*. An order includes animals that are usually similar in their appearance and way of life and are clearly distinct as a group from members of other orders. For example, the members of the order Carnivora – cats, dogs, bears, hyenas and seals – are clearly different from members of the order Primates, which includes the monkeys, apes and man.

Orders are grouped into classes, and classes into phyla (singular 'phylum'). The class Mammalia includes all the warm-blooded animals that have hair and feed their young on milk produced by the mother. In contrast the class Aves (birds) includes warm-blooded animals with feathers and wings. These two classes together with the reptiles, amphibians and fishes and a few strange animals such as seasquirts are grouped in the phylum Chordata. Members of different phyla are usually distinguished on the basis of major differences, particularly in their method of locomotion. At some time in their

species	genus	family	order
Cervus elaphus	*Cervus*	Cervidae	Artiodactyla

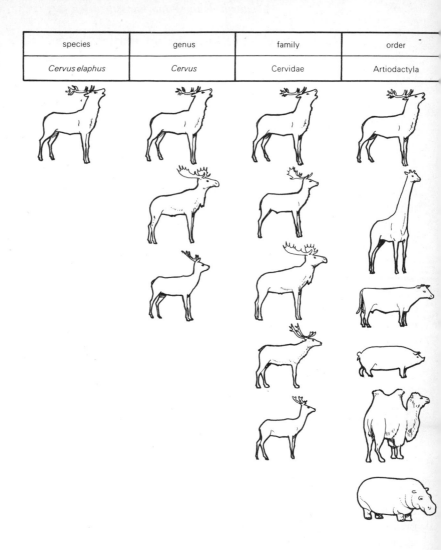

This diagram shows the place of one animal – the red deer, **Cervus elaphus** *– in the various animal groupings made by scientists*

class	phylum	kingdom
Mammalia	Chordata	Animalia

lives all chordates have a stiffening rod along their backs. This allows them to swim by moving the body from side to side, and from this they developed the ability to walk on four legs. In contrast, the jointed limbs of insects, crabs and spiders (phylum Arthropoda) allow them to creep along in a characteristic way.

Some parts of classification are more difficult to apply to fossils than to living animals. For example, members of fossil species cannot be interfertile as they may have been dead for millions of years. Therefore fossil species are usually defined on the basis of minor differences from other closely related species. Beyond this, interpretation of the features of fossils usually allows them to be classified with a good degree of confidence.

The division of animals into phyla reflects their basic relationships, and these divisions are used in most of this book and in most other books on fossil and living animals. It is therefore useful to have an understanding of these groups as it will help you to identify fossil as well as living animals more easily.

2 *Identifying Fossils*

As soon as you have begun to collect fossils you will want to find out exactly what they are. Unfortunately there is no single book or even small group of books that will allow you to identify quickly all your specimens. Descriptions of the hundreds of thousands of fossil species known are scattered through many hundreds of scientific journals. There is, however, no need to be disheartened by this, because few professional geologists can quickly refer every specimen that they find to its correct species. To do this usually requires specialist knowledge of the fossil group concerned, an extensive library and the availability of a collection of previously identified specimens for comparison. For this reason you should attempt to identify your specimens to the genus level only.

The brief descriptions that follow should enable you to place most of your fossils in their groups. Plants will probably give you most difficulty. At the end of this book, on pp. 157–8, we have listed several useful and easily obtainable books that will help you with identifications. The three British Museum (Natural History) handbooks of *British Palaeozoic Fossils*, *British Mesozoic Fossils* and *British Caenozoic Fossils* are particularly valuable aids to identification. Between them they illustrate nearly 1,200 species and most of these are common forms.

You can supplement information from books by looking at the exhibits in museums, since you will often be able to identify your specimens by direct comparison with the fossils on display. However, you should remember that the names of fossils change fairly often as a result of reidentification, so you may find that the labels in the museum are not up to date or do not agree with the names that you have got from your guide books. If you need further information you should contact the museum curator, who may be able to help you or can put you in touch with someone who will.

VERTEBRATES

Vertebrates or animals with backbones include all the familiar larger animals of the world such as birds, reptiles, mammals, fishes and amphibians. Fossil vertebrates are the most popular and exciting of all fossils, and every day thousands of people visit larger museums to see large fossil fishes, dinosaurs and flying reptiles. The hard parts of invertebrates usually consist of a shelly or chitinous coating that covers the soft parts of the animal. In contrast the hard parts of a vertebrate are usually internal and are covered by soft parts.

Vertebrates are rare as fossils and there are several reasons for this. Firstly, many vertebrates live on land and are therefore less likely to be fossilized (see p. 13). Secondly, even in the sea vertebrates are less common than invertebrates. Thirdly, the bones of vertebrates are usually only loosely joined to each other, so that they may be scattered more easily or eaten by scavengers. Fourthly, there are no vertebrates that fix themselves to one spot for their adult life, whereas many of the commonest invertebrate fossils, such as bivalve molluscs, corals, bryozoans, sponges and sea lilies, represent animals that as adults are permanently fixed.

This is not to say that you will not find vertebrate fossils; but skeletons are very rare and you may have to be satisfied with collecting isolated bones, scales or teeth. Also vertebrate remains often require special collecting techniques. In many cases fossil vertebrates occur in deposits where invertebrates are rare, and similarly fossil vertebrates are often rare in places where invertebrates are common.

Of the vertebrate groups, fishes, reptiles and mammals are important as fossils. Amphibian remains occur in Palaeozoic and later rocks, and you may find frog remains in Pleistocene deposits. Birds are rare as fossils because their bones are very delicate and therefore rarely fossilize; bird bones are sometimes found, however, at Sheppey (see p. 80) and in other parts of south-east England.

Fossil fishes

Fossil fishes are here described in three groupings – the armoured fishes, the sharks and rays, and the bony fishes. The last two are natural groups, but the first includes several groups of Palaeozoic fishes that all had heavy bony armour over their bodies.

Armoured fishes were important in the Palaeozoic. All of them had a bony body covering, which may have protected them but also meant that they were poor swimmers. In many of the early armoured fishes the head was covered by a head shield which consisted of plates of bone joined together. During fossilization these bony plates often became separated, so that only isolated plates are usually found. We found plates of Palaeozoic fishes at Lydney (see p. 133).

Sharks and rays have skeletons that are made of cartilage or gristle. This is soft and rarely fossilizes, but the teeth of these fishes are probably the most common remains of any fossil vertebrates. The teeth of sharks begin growing on the inner edge of the jaws and they slowly move into position, replacing earlier teeth. As they become worn or broken the teeth move sideways and are replaced by new teeth. Eventually the old teeth fall out. As teeth are continually being produced, a single shark can have several hundred in its lifetime and these can accumulate in deposits. Vast numbers of sharks' teeth can be collected from Eocene sands in parts of Kent, and good specimens can be collected on the beach at Sheppey (p.

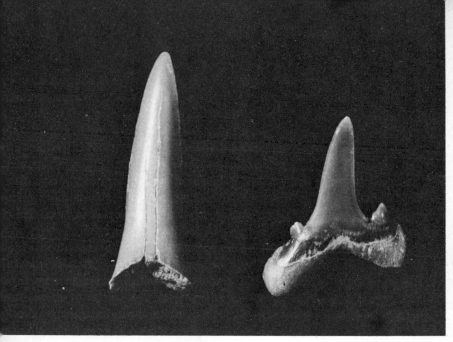

Sharks' teeth from Herne Bay, Kent

83) and Herne Bay (p. 79). A well-preserved shark's tooth consists of a shiny pointed region and a forked region with a dull surface. The dull region is the root, and in life it is attached to the jaw, while the pointed region is the crown, used for biting. Important features of sharks' teeth are their size, overall shape and the shape of the crown. Additional points may be present flanking the main point.

Since most sharks are active hunters, their sharp teeth are important for gripping their prey. But some sharks, such as *Acrodus*, and rays feed on molluscs and crustaceans. This diet requires strong blunt teeth that can be used to crush the food. The teeth of rays are flattened and plate-like. We did not find any of these teeth, but they occur at Barton in Hampshire and in the London Clay or Blackheath Beds at, for example, Abbey Wood in Kent.

Some sharks also had long spines that grew in front of their fins. These 'fin spines' occurred in *Acrodus* from Aust (p. 131). The **bony fishes**, such as the herrings, trout, mackerel and cod, have skeletons that are made of bone. This is a far harder substance than cartilage and as a result it fossilizes more easily. Bony fishes are extremely common in modern seas, and with over 20,000 living species they outnumber all other vertebrate species combined. This great diversity is also found in fossil bony fishes, and as a result many of them are very hard to identify. If, however, you find a good specimen, most museums will be glad to identify it for you. Good specimens of bony fishes are rare. In Britain they can be found at Sheppey and Charmouth. We found scales or bones of bony fishes at Whitecliff Bay (p. 112) and Charmouth (p. 89).

Slabs of rock or nodules containing fossil fishes are often seen on sale in rock shops or curio shops and are usually quite expensive. Most of these fishes come from Brazil, Italy or Lebanon.

Reptiles

Fossil reptiles are not uncommon in Britain, and if you are in the right area bone fragments are easily collected. The reptiles enjoyed their greatest success during Mesozoic times, when the dinosaurs were the dominant larger land animals, while the pterosaurs were important flying animals, and in the seas and freshwater there were turtles, crocodiles, ichthyosaurs and plesiosaurs. The reptiles declined at the end of the Mesozoic, but the crocodiles and turtles were common throughout the Cainozoic (Tertiary), and lizards and snakes are encountered as fossils in some deposits that are mainly of freshwater origin.

Dinosaur remains occur in parts of south-east England, in Oxfordshire and on the Isle of Wight. Fairly complete remains have been found on the Isle of Wight, but they are very rare. However, dinosaur bones are often found along the cliffs from Compton Bay to Atherfield Point. The remains of flying reptiles or pterosaurs also occur on the Isle of Wight, and the first pterosaur remains ever discovered were found at Lyme Regis.

Reptile bone in a pebble from Charmouth, Dorset

As may be expected the commonest fossil reptiles are those which lived in water, and Britain is particularly rich in the remains of plesiosaurs and ichthyosaurs.

Plesiosaurs had a way of life that was probably similar to that of the living seals. They had short bodies, very long slender necks and long tails. Their four flippers were well developed and they used them to 'row' themselves through the seas. Bones of plesiosaurs were found at Aust (p. 130). Remains of plesiosaurs also occur along the south coast of Dorset, near Street in Somerset and in the brickpits of Leicestershire and Bedfordshire.

Ichthyosaurs looked very like fishes, with well-developed fish-like tails and even a high dorsal fin like that of a shark. Ichthyosaurs lived very like whales and dolphins. Their limbs were completely like fins and were used only for steering, as the ichthyosaurs never needed to come on land because they could give birth to live young in the sea. Remains of ichthyosaurs occur in most places where plesiosaurs are found, and in the last century many ichthyosaurs were collected from near Lyme Regis. We are not certain that we collected remains of ichthyosaurs, but a single pebble containing bone from Charmouth (p. 89) was either from an ichthyosaur or a plesiosaur.

Crocodiles first occur in the Triassic, but their remains are not common in Mesozoic rocks. In Cainozoic rocks, however, the remains of crocodiles and turtles are often the commonest vertebrate fossils represented. Living crocodiles have a row of bony plates that extend along their backs from the top of the skull to the tail. These plates are usually several centimetres across, and each one has a deeply pitted upper surface that is characteristic and allows even small bone fragments to be identified as crocodile. Teeth of crocodiles are also common in many areas. These teeth are conical, with fine ridges and grooves running along their length. The lower ends of crocodile teeth are hollow, with a conical space. Bones of crocodiles also occur, but they may be difficult to distinguish from those of turtles.

Turtles are usually common in the same places as crocodiles. The parts of turtles that are most often found are fragments of their shells. These are never pitted as strongly as the plates of crocodiles, and they are usually thicker than the bony plates of some fishes. The plates of freshwater turtles have a pattern on one surface, while the plates of marine turtles are smooth on both surfaces. Remains of turtles and crocodiles are common in the Purbeck Beds (Jurassic/Cretaceous) at Swanage in Dorset and in the Hamstead Beds (Oligocene) at Yarmouth on the Isle of Wight.

Mammals

Mammals originated in the Mesozoic. Mammals of this age have been found in parts of Wales, Somerset, Oxfordshire,

Dorset, Sussex and Western Scotland, but these remains are very rare. They need specialized equipment for their collection and very specialized knowledge for their identification.

The Cainozoic (or Tertiary) is also known as the Age of Mammals, because they were dominant on land. Remains of Cainozoic mammals occur in Kent, Hampshire and on the Isle of Wight (p. 109). Fossil mammals are identified from their cheek teeth, and in many cases the teeth are the only parts that are preserved. Teeth are the hardest part of the mammal skeleton: they are compact and break less easily than the bones. Cainozoic mammals are not easy to find anywhere in Britain, but we were able to collect some teeth from Headon Hill near Yarmouth on the Isle of Wight (p. 119).

INVERTEBRATES

This group of animals includes snails, worms, sea urchins, sponges, corals and insects. It is a convenient grouping of animals without backbones or indeed any other bones. Invertebrates are far commoner than vertebrates both as living animals and as fossils. The insects are more abundant and varied than all other animals combined; and the molluscs, the group that includes snails, cockles and squids, form the commonest fossil group of all. In Britain there are few areas where you can collect vertebrates, but fossil invertebrates can be found in most areas that contain sedimentary rocks.

Protozoa

Members of the phylum Protozoa are single-celled animals and most of them are very small. Living protozoans include the amoeba. The most important fossil protozoans belong to the groups Foraminifera and Radiolaria.

Foraminiferans, or forams for short, are usually so small

Nummulite from Whitecliff Bay, Isle of Wight

they need special collecting techniques and can only be identified using a microscope. However, forams are extremely important in geological work, especially in studies of Upper
Mesozoic and Cainozoic rocks. They are very common in most
sedimentary rocks that were deposited in the sea, and as they
are so small a good sample can be extracted even from the
small amounts of rock that are recovered from the borehole
formed when drilling for oil. Some of the largest forams are
called nummulites. These usually have flattened disc-shaped
shells, made up internally of a large number of chambers.
They can occur in vast numbers in some Palaeocene and
Eocene rocks. We collected nummulites up to a centimetre in
diameter at Whitecliff Bay on the Isle of Wight (see p. 113).
The identification of most forams is extremely difficult, even
for an expert, but you may be able to identify them if only a
few species are known from the locality where you were collecting.

Radiolarians are almost all less than 0·1 mm across and therefore they cannot be studied without a microscope. Today radiolarians live in very deep seas. They have shells made of silica, and their remains accumulate to form deposits that are called 'radiolarian oozes'. Usually these form only at depths greater than 2,300 m, and the discovery of rocks formed from radiolarians suggests deposition in very deep water. Rocks formed from radiolarians occur in Britain in the Lower Carboniferous cherts of North Devon.

Sponges

Fossil sponges first occur in Precambrian rocks, and the sponges have a longer fossil record than any other animal group. Many sponges have changed only gradually through time, so they are of little use for correlation. The identification of sponges is usually based on a study of polished sections and is very difficult. Fortunately, however, some sponges have characteristic shapes which allow their identification.

There are three main groups. The sponges most familiar to us are called **horny sponges**. They have soft skeletons that are made of a material called 'spongin'. Fossils of these sponges do not exist.

The skeleton of a sponge consists of small, many-rayed units that are called spicules. These may be dispersed through the tissues of the sponge or they may join together to form a rigid support. The **glass sponges** have siliceous skeletons. They are often vase-shaped with long narrow bodies. *Ventriculites* (p. 75) is a typical fossil glass sponge.

Calcareous or **limy sponges** have skeletons made of calcite; they are very irregular in shape, but many of them are rounded or spherical. *Porosphaera* (p. 112) is a calcisponge.

Some sponges, such as *Cliona*, live by boring into rocks or shells. Fossils of *Cliona* may occur as small swellings on the surfaces of rocks or shells, or they may be seen as trace fossils consisting of the small slot-shaped holes made by the sponge (p. 74). Fossil *Cliona* burrows were collected at Portland Bill (p. 104) from the Pleistocene raised beach deposits, while Recent borings were found in the chalk pebbles at Botany Bay (p. 74) and are common on many beaches.

Many fossil sponges have pores on their outer surfaces, and all have a large opening or *cloaca* at one end and an attachment point at the other. Sponges may be confused with some corals or bryozoans or with one group (rudists) of molluscs, but local guides and faunal lists should help to prevent any confusion.

Corals

Corals, sea anemones and jellyfishes are grouped in the phylum Coelenterata. Corals produce a skeleton made of calcite, which can be fossilized easily, and as a result they are by far the most important fossil coelenterates.

Cliona *burrows from the raised beach at Portland Bill*

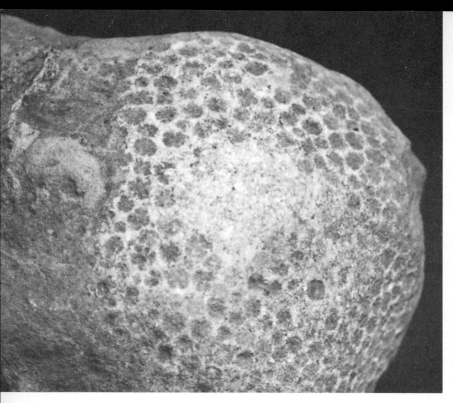

Holocystis, *a coral, from Atherfield Point, Isle of Wight*

An individual of a coral colony or a solitary individual is called a corallite, and each corallite consists of a calcareous tube or cone containing structures that are important for the identification of the coral. Most corals have many pores on their surfaces. These look like tiny stars, with small ridges or lines projecting into their centres. These ridges are the ends of tiny vertical plates that are called *septa*. You may need to use your hand lens to see these septa, but some corals have large pores and easily visible septa. Solitary corals may be easily confused with sponges, but details of their structure should help you to distinguish members of the two groups.

There are three main groups of corals that are important as fossils.

Rugose corals often have wrinkled surfaces, from which

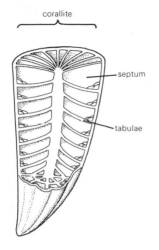

corallite

septum

tabulae

A cross-section through a coral showing the scientific terms that are used to describe its features

the group gets its name. They may be common in Ordovician to Permian rocks but are unknown from Mesozoic rocks. This is very useful, as many rugose corals are similar in appearance to scleractine corals, which do not occur in rocks older than the Triassic, so if you know the age of the rocks in which you are collecting you can distinguish between the two groups simply on an age basis. Rugose corals may be colonial or solitary. Most of them have clearly defined septa, and they may also have horizontal plates or *tabulae*. Many solitary rugose corals are conical, with their attachment at the point and their upper surface concave with many septa running inwards to a deep central depression or hole.

Tabulate corals are always colonial, and they occur almost entirely in Palaeozoic rocks. Many tabulate corals have a branching or root-like growth form, but some may grow as clumps. Tabulate corals are characterized by the presence of horizontal tabulae, and if septa are present they are always very short.

Scleractine corals range from the Triassic to Recent. They may be solitary or colonial, and they have all types of growth form. Septa are usually prominent, but tabulae may also be present.

Identification of corals is often based on studies of thin sections. Details of the septa are particularly important. Many solitary corals have characteristic shapes and these may sometimes be identified easily, but the identification of colonial corals is often a job for the expert.

Bryozoans

Bryozoans are also called moss animals. This is an accurate description of these tiny, often delicate colonial animals. They usually grow as small, irregular shaped patches encrusting rocks or shells, but some are like small delicate ferns. The surface of a bryozoan is covered with small pores that look like a pattern of pin-pricks, and because of this they may be confused with corals. Bryozoans may be found on the surfaces of other fossils, or they may be washed out of clays. They can also be collected from the weathered surfaces of limestones and chalk.

Brachiopods

Brachiopods are one of the most important groups of fossils and are abundant in many Palaeozoic and Mesozoic rocks. They declined slightly in the Cainozoic but are still common in marine faunas throughout the world.

Brachiopods have shells consisting of two valves. The front part of the shell is usually pointed and is called the *beak*. Just behind the beak is the region where the two valves articulate. This is called the *hinge line* and may be very short (for example, *Rhynchonella*), or it may be the widest part of the shell. The region between the hinge line and the beak is called the *interarea*. The presence and size of the interarea may be important for identification. One valve always has a larger beak than the other, and the valve carrying the larger beak is called

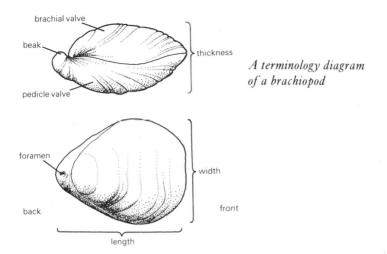

A terminology diagram of a brachiopod

the *pedicle* valve, while the other is called the *brachial* valve. It is important to distinguish the two valves before attempting to identify a brachiopod. The thickness of a brachiopod is the vertical distance through the pedicle and brachial valves. The length is the distance from the beak to the back of the shell, and the width is the greatest distance across the shell.

There are two major groups of brachiopods – the inarticulate and the articulate kinds.

Inarticulate brachiopods are generally uncommon as fossils. They are known as 'inarticulate' because the hinge between the two valves is very poorly developed, so that the valves do not articulate together. As a result the valves are not usually found joined together. Some inarticulate brachiopods are found attached to other fossils, while others, such as *Lingula*, have glossy shells and look like fingernails.

Articulate brachiopods are usually found with the valves joined together. There are eight orders of articulates, and six of these are important as fossils. The orthids, Pentamerids and spirifers are limited to the Palaeozoic, and the strophomenids are important only in the Palaeozoic. The two orders Terebratulida and Rhynchonellida range from the Ordovician to Recent, but are important mainly in the Mesozoic and Cainozoic.

Orthids have shells that are almost circular to elliptical. Both valves are convex and often carry a pattern of ribs. Interareas are present on both valves, and the hinge line is usually long. *Orthis* is a typical orthid, and it contrasts with the pentamerid *Conchidium*, which has a short hinge line. Members of these two Palaeozoic orders can therefore be distinguished by the length of their hinge lines; and they can be distinguished from strophomenids because they have shells with both valves convex, whereas strophomenids usually have one valve convex and the other concave. When identifying strophomenids it is important to find which valve is concave. To do this look carefully at the beak. Unfortunately there are no clearly defined external features to distinguish spirifers. Members of this group are very variable: they may have long or short hinge lines, interareas may be clearly visible or they may be difficult to find, and the shell may have a pattern of ribs or it may be quite smooth.

We found terebratulids and rhynchonellids at Crickley Hill (p. 141), Gilbert's Grave (p. 139) and Atherfield (p. 122). Terebratulids have interareas on their pedicle valves only; they

Sellithyris and Sulcirhynchia, *brachiopods, from Atherfield*

usually have smooth shells, and there is usually a small hole or *foramen* on the beak of the pedicle valve. In contrast, rhynchonellids have interareas that are small or not visible, and their shell surface usually carries strong ridges. It is therefore usually fairly easy to distinguish them from terebratulids.

Molluscs

Living molluscs include snails, cockles and cuttlefish. The group is divided into three main sub-groups – the bivalve molluscs, cephalopods and gastropods. Molluscs are by far the commonest larger fossils in Mesozoic and Cainozoic rocks. The abundance and variety of molluscs make them a very difficult group for the amateur collector and for the expert. To identify molluscs use the faunal lists from the site where you were collecting, and remember that even in the same species there can be great variation during growth, and that even in the adult stages you may not get exact agreement between your specimen and pictures of the species.

Bivalve molluscs include all the familiar living shells that consist of two valves: for example, clams, cockles, oysters, mussels, scallops and razor shells.

The brachiopods are the only fossil group that can be confused with bivalve molluscs, but with a little care members of the two groups can be distinguished very easily. In brachiopods the pedicle and brachial valves are usually different. They differ in curvature and size, and their beaks are always different. In contrast the two valves of a bivalve mollusc are usually similar, being mirror images of each other. Exceptions to this are the oysters and many scallops. Each valve of a brachiopod can be divided along its mid-line and the two halves are then mirror images, that is, the valves are bilaterally symmetrical. In contrast the valve of a bivalve mollusc is not symmetrical and therefore cannot be divided into similar halves.

A nest of Corbula, *a small bivalve mollusc, from Bishopstone Glen, Kent*

Bivalve molluscs first occur in the Ordovician and they are common in some Carboniferous rocks. Bivalves become increasingly common in rocks of Mesozoic and Cainozoic age, and the group may now be at the peak of its development.

Before describing or identifying a bivalve mollusc it should be positioned correctly, and this positioning is very different from that of a brachiopod. The correct position and dimensions of a bivalve mollusc shell are shown opposite.

The two valves articulate in the region near the beak; this is again called the *hinge*, and if it is elongated it is called a *hinge line*. On the hinge near the beak there may be strong swellings or projections of the shell; these are called *hinge teeth*. Those

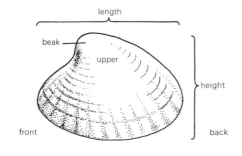

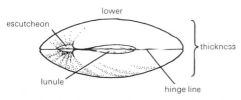

A terminology diagram of a bivalve mollusc

near the beak are called *cardinal teeth* and those away from the beak are called *lateral teeth*. The shell near the hinge line may be hollow or flattened. A hollowing in front of the beak is called a *lunule*, while one behind is called an *escutcheon*.

Inside the valve there may be flattened or depressed regions that are often polished. These lie near the front and back of the shell and are *muscle scars*, which indicate the areas of attachment of the strong muscles that are used to open and close the valves. When the valves are pressed together there may still be openings between the edges of the opposing valves. Such an opening is called a *gape*, and in life it allows the fleshy foot of the bivalve to be kept outside the shell even when the valves are closed.

The squid, cuttle fish and octopus are living **cephalopods**. They are a highly successful group and are common in most seas of the world. The commonest remains of Recent cephalopods found in Britain are the cuttle bones that are washed up on most beaches. Cephalopods never occur in freshwater.

Nautiloid cephalopods are represented today by only a single genus – *Nautilus* – which lives in the Pacific. *Nautilus* has a large coiled shell that consists of many separate chambers. The

outer or last chamber holds the soft parts of the animal and the
other chambers are filled with gas. As a result *Nautilus* swims
with its soft parts and last chamber hanging below the gas-
filled part of the shell.

Fossil nautiloids are common in some early Palaeozoic rocks,
but the group declined steadily throughout the late Palaeozoic
and the Mesozoic. Some nautiloids have strongly coiled shells,
but others have straight slender, short thick, or loosely coiled
shells.

The chambers of the nautiloid shell are divided from each
other by thin shelly walls known as *septa*. These meet the shell
along *suture lines*, which can only be seen when the shell is
removed so that the internal mould is visible. In nautiloids the
suture lines are characteristically simple, whereas in most ammonites (see p. 88) they are more complex.

Belemnites are also very common in rocks of Mesozoic age.
They declined very rapidly during Cretaceous times, and
although a few survived into the early Cainozoic you are
unlikely to find these; so the presence of belemnites in a rock

Belemnites at Seatown, Dorset

means that it is almost certainly of Mesozoic age. Belemnites are shaped like bullets, and this part of the belemnite is called the *rostrum*. In life the rostrum formed the back of the animal with the point backwards. Grooves or ridges may be present on the rostrum. Well-preserved belemnites have a thin-walled hollow region at the front of the rostrum. This is called the *phragmocone* and it contained some of the soft parts of the animal. When the phragmocone is present it is usually crushed but its thin walls are then visible.

Ammonites are by far the commonest and most familiar fossil cephalopods. Most ammonites have tightly coiled flattened shells that might be confused with a few gastropods, but are otherwise unlike any other fossils. Some ammonites, however, have loosely coiled shells that may even be straight or conical, but as these kinds are exceptional they are usually mentioned in guides and descriptions of sites where they occur.

Most fossil ammonites consist of internal moulds. We collected specimens from Folkestone (p. 88) in which the shell was preserved, but this is unusual and most of the important features used to identify ammonites are found on the internal moulds.

The hollow region on the side of the ammonite is called the *umbilicus*. The opening of the shell is called the *aperture*. Each turn of the shell is called a *whorl*, and the whorls join along a *seam*. The outer edge of the whorl is called the *venter* and a ridge in this region is called a *keel*.

The ammonite shell is divided internally into many separate chambers that are separated by *septa*, and, as in nautiloids, the line where the septum meets the shell is called a *suture*. The pattern that the suture makes on the wall of the shell is very important for identifying ammonites. The suture pattern is often complex and may look like fern leaves or frosting, but in other cases the pattern may be simple, consisting only of a wavy line around the shell. Suture patterns generally became more complex throughout the Mesozoic, but during the

Small Androgynoceras *ammonites from Charmouth, Dorset*

Cretaceous some ammonites appeared with very simple suture patterns that resemble those found in Palaeozoic forms.

If the shell of the ammonite is preserved it may be necessary to scrape some of it away to expose the suture lines. Polishing of the surface of an internal mould may make the suture more visible, and in heavily worn specimens, such as those occurring at Folkestone, the suture may show very clearly. Some ammonites break along their suture lines, and this reveals the suture and the complete septum.

Most **gastropods** are like snails, having tightly coiled shells, and the group includes land snails, whelks, winkles and the land and marine slugs. Marine snails are the commonest fossil

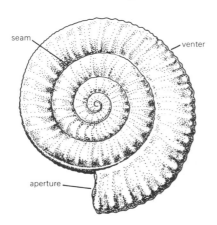

seam

venter

aperture

A terminology diagram of an ammonite shell

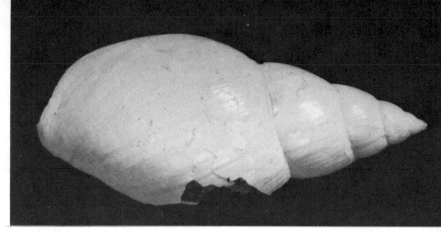

Lymnaea, *a gastropod, from Headon Hill, Isle of Wight*

gastropods, and several features of the shell are important for identification.

The opening of the shell is called the *aperture*, and each complete coil of the shell is called a *whorl*. The *last whorl* runs from the aperture once around the shell and is always the largest of the whorls. The rest of the shell is called the *spire*. The junction between the whorls is called the *suture*. The aperture may be rounded or its lower (front) part may be extended to form a *canal*.

Patterns of sculpturing on the shell, the type of coiling, shape of the shell and details of the aperture may be important for identification.

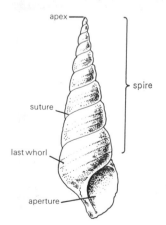

A terminology diagram of a gastropod

Gastropods occur in rocks of early Palaeozoic age, and the group has always been important. They increase steadily in diversity and abundance throughout the Mesozoic, and in Lower and Middle Cainozoic rocks they reach the peak of their diversity, with a host of different species occurring at sites such as the Eocene of Whitecliff Bay on the Isle of Wight (p. 109).

Most gastropod fossils occur in marine deposits but they are also common in some freshwater deposits (for example the Eocene beds of Headon Hill on the Isle of Wight). Gastropods first became able to live on land in Carboniferous times about 300 million years ago, but fossil land gastropods are known mainly from Cainozoic rocks. Land and freshwater gastropods usually have thin, brittle shells, whereas those of marine gastropods may be very thick.

The **scaphopods** form a small class of molluscs. Members of the group have long tubular shells that are open at both ends. Fine ridges run along the shell in some genera, including *Dentalium*, which we found at Folkestone (p. 88). There may also be very weak growth lines running around the shell.

Dentalium, a scaphopod, from Folkestone, Kent

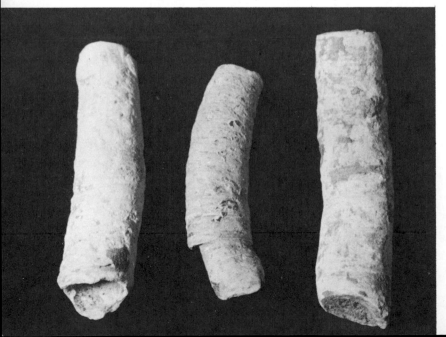

Worms

Worms such as the earthworm, ragworm and lug worm are soft-bodied animals that rarely fossilize. However, some worms such as *Serpula* and *Spirorbis* produce hard tubes in which they live. These chalky cases enclose the soft parts of the animal when it is alive and they often fossilize. Worms are also important because they burrow extensively and their fossilized burrows are important as trace fossils. Many fossil burrows, such as *Chondrites*, have names that are used only for the burrow, and in many instances the soft-bodied animals that produced these burrows remain unknown.

Trace fossils such as these may be restricted to small areas, and they will be mentioned in local guides.

Chondrites, *worm burrows, from Charmouth, Dorset*

Arthropods

The arthropods are the most abundant living invertebrates, and there are more species of arthropods than all other animals combined. The group includes lobster, crabs, shrimps and prawns as well as the millipedes, centipedes, spiders, scorpions and insects.

Arthropods have jointed legs and their bodies have a hard outer coating. This is made of a substance called *chitin* and it protects the arthropod as well as providing a rigid surface against which its muscles can pull. In many marine arthropods the chitinous body coating is strengthened by becoming impregnated with calcium salts. This gives a thick hard coating like that of crabs and lobsters. The presence of this hard coating and jointed limbs means that arthropods can only grow by moulting their body coating and increasing in size while the new coating is still soft. Moulting occurs at regular intervals, and the moulted coating may retain the form of the animal that

Meyeria, *a small lobster, from Atherfield, Isle of Wight. Parasitic molluscs can be seen attached to it*

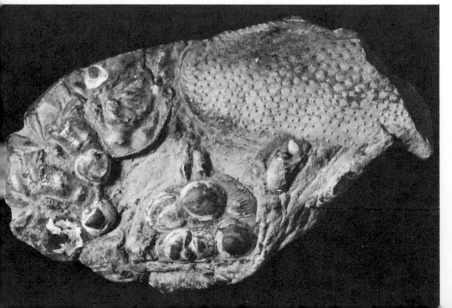

produced it. This is important because many fossil arthropods consist only of the moulted coat.

The **trilobites** are the most important of the larger fossil arthropods. They occur only in Palaeozoic rocks, and in many Cambrian and Ordovician rocks they are the commonest fossils. The trilobite body is divided into three parts by a pair of furrows that run along the full length of the body, and from this the name 'trilobite' is derived. The central part defined by these furrows is called the *axis* and the two side parts are called *pleural lobes*. The body is also divided into a head (*cephalon*), a thorax and a tail (*pygidium*). Important features of trilobites are their size and general shape; eyes may be present. The grooves on the axis of the head and the shape of the back corners of the head – the *genal* regions – may also be important for identification. *Genal spines* of varying length may be present, or the genal regions may simply be rounded. The tail may consist of many or a few segments, or it may be fused into a solid plate; also a series of small spines or a single large tail spine may be developed.

Crabs, lobsters, prawns, shrimps and barnacles belong to the Crustacea, and throughout the Mesozoic and Cainozoic **crustaceans** have been important marine and freshwater arthropods. However, they have never been successful on land,

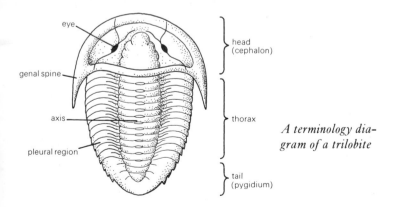

A terminology diagram of a trilobite

where they are represented only by a few land crabs and the wood lice.

Medium-sized and large crustaceans are not common as fossils, and most of them can be easily placed with their group as they are very similar to their living relatives. It is unusual to find complete crustaceans: fragments of limbs or the body coating are the commonest remains. We found crustacean fragments at Crickley Hill (p. 141), and we found almost complete lobsters at Atherfield on the Isle of Wight (p. 125).

Barnacles look very like some molluscs, but their hard outer shell contains a small animal that is similar in general appearance to a small shrimp. Barnacles occur attached to other fossils and may be common in some Pliocene and Pleistocene deposits.

Ostracods are tiny crustaceans, usually less than 5 mm long. Most ostracods have bivalved shells, which means that they look very like tiny bivalved molluscs, but under a strong lens or microscope the valves may be seen to have strange sculpturing which is usually unlike that of molluscs.

Ostracods and forams are very important in stratigraphical work, as they are small and can be found in most sediments. Forams are particularly important in marine sediments but they do not occur in freshwater deposits. Ostracods, however, occur in both kinds of sediments, which makes them especially important.

We found ostracods at Atherfield (p. 122). Huge numbers of them were lying along the bedding planes of the rocks, which gave the appearance of a grey pitting when the rock was split. In most areas ostracods can only be collected by sieving the sediments.

Today **insects** are extremely common and varied. They live in most land areas and survive under extreme weather conditions, though most have delicate bodies with only a thin coating of *chitin*. In view of their abundance in the modern world insects are surprisingly rare as fossils. This is probably accounted for by their delicate bodies and because there are relatively few aquatic insects. Identification of most living insects is difficult enough, and the identification of fossil insects is often a job for the expert. The hardest part of any insects are the wing cases of larger beetles, and these occur in many Pleistocene freshwater deposits.

In contrast the most delicate parts of insects are their wings, which are folded and protected in beetles but are permanently exposed in insects like crane flies and dragon flies. Insect wings can be found at Gurnard Bay on the Isle of Wight, where complete wings are rare but wing fragments are fairly easy to collect.

Sea urchins and starfishes

Sea urchins, starfishes and their relatives belong to an important phylum called the Echinodermata. They differ from other animals because their bodies have a star-shaped pattern based on five rays. This is obvious in the starfishes, and can be seen on the top of many sea urchins as five raised or depressed areas or five petal-like regions.

Starfishes are uncommon as fossils, and in areas where they do occur they will be well described in the local guide. If you are collecting starfishes you may encounter only a few fragments of the body or pieces of an arm. Identification is usually not difficult with the help of the guide.

Sea urchins or *echinoids* range in shape from shallow discs to nearly spherical forms. Well-preserved fossils may be covered with long spines, but these are usually missing, so that only the main part or *test* is preserved. This test consists of many fused plates, and the form of these may be important for identification.

A sea urchin has two main body openings – the mouth and the anus. The mouth is always on the lower flattened surface of the animal, but the anus varies in position. Regular echinoids have the anus at the centre of their upper surface and the

Holectypus, *a sea urchin, from Crickley Hill*

mouth directly below it at the centre of the lower surface.
They are circular in outline, and many of them have swellings
or tubercles arranged in bands over the test. Regular echinoids
first became common in Mesozoic rocks. They survived
through the Cainozoic and are still common marine animals.

Irregular echinoids are usually more common as fossils. The
anus of irregular echinoids·is shifted towards the back or even
underneath the animal. For example, in *Micraster* the anus is
on the side of the pointed back end of the animal, while in
Echinocorys it is underneath the test. At the same time the
mouth is rarely at the centre of the lower surface and may be
shifted to the front. Many irregular echinoids have an outline
that is lengthened in the plane of the mouth and anus. A

common exception to this is *Conulus*, which has a circular outline, but in *Conulus* the anus is on the lower surface, so it is easily identifiable as an irregular form.

Although they may be very common there are usually only a few genera of echinoids at any one site. This means that identification is often easy with the help of a local guide.

Sea lilies or crinoids look very like plants. Well-preserved specimens have a stem, a small body and many long delicate arms that may have a feathery or fern-like appearance. The stem is made up from small disc-like plates that are joined together in columns. These plates are called *ossicles*, and sometimes they may be used to identify the crinoid. For example, *Pentacrinus* has star-shaped ossicles. The crinoid body is made up of many small plates that are joined together. Complete specimens are rare, but fragmentary remains of crinoids are common and sometimes they may form the bulk of a rock. Identification of crinoids is usually based on details of the body plates, but the size of the body and a simple count of the plates may help with identification, while details of the arm bases will also help.

Fragments of Pentacrinus, *sea-lily stems, from Charmouth, Dorset*

Graptolites

Graptolites are a very important group of Palaeozoic fossils, and even though we did not collect any, a brief description of them is necessary.

Graptolites frequently occur in shales or slates, and they may be found by splitting the rocks along the bedding planes. If this is done the black glossy graptolites may be seen as they contrast with the matt surface of the rock and look like pencil lines on black paper.

Each graptolite was a colony of small animals that lived in cup-like cases called *thecae* (singular 'theca'). These cases joined in rows to form the *arms* of the graptolite. In well-preserved specimens the openings of the cups may be visible as saw-like serrations along one or both edges of the arms.

Several technical terms are used in descriptions of graptolites. The arms are called *stipes*, and the point where the main stipes meet is called the *sicula*. A stem or *nema* is produced from the sicula. Cross-connections between stipes are called *dissepiments*.

For identifying graptolites the number and position of the arms is important, but during preservation the graptolites may have landed in many different positions, so that members of the same species can look very different. To overcome this it may be necessary to interpret your fossils carefully and to work with as many specimens as possible.

Graptolites are common in many areas. In Britain they can be found at Abereiddy Bay near St David's, Dyfed, in the Ordovician Shales; and also in the Ordovician and Lower Silurian Shales at Builth.

PLANTS

Fossil plants are very common in some areas, but in general they are rarer than fossil animals. Plants are also less varied than animals, and local guides will often contain sufficient information for you to identify any fossil plants you find. The earliest plants that are common as fossils occur in the Coal Measures and are Carboniferous in age. During Coal Measure times large coastal areas of Britain were covered with dense forests and marshes. These forests were inhabited by insects, fishes and large amphibians, and they contained a wealth of plants of many different types and species. The plants from these forests became fossilized as coal, with the result that the coal-mining areas of Britain are now rich in fossil plants. These can be collected from the tip heaps of coalmines, and some of them are beautiful specimens. Leaf impressions, pieces of tree trunk or branches and fruit can all be found. Most of these are in the form of black carbonized fossils, and the best specimens are obtained by splitting blocks of shale, nodules or the clayey rocks that occur on tip heaps.

The fossil plants of the Coal Measures include relatives of the living club mosses. These fossils are called scale trees because the bark of their trunks and branches carried many small scales which left diamond-shaped scars when they broke off. These scars may be arranged spirally, as in *Lepidodendron*, or vertically, as in *Sigillaria*.

Perhaps the most well-known Coal Measure fossils are the forms which are related to the living horsetails. These had jointed stems that were hollow, and internal moulds of the stems carry vertical grooves in the genus *Calamites*. Small star-like leaves called *Annularia* probably belonged to Horsetails like *Calamites*. Fossil ferns also occur in the Coal Measures, and these look like the fronds of living ferns. Horsetails and ferns reproduce by using spores, but many fern-like plants represented in the Coal Measures have been found to reproduce by means of seeds. These plants are called seed ferns, but their remains are difficult to distinguish from those of true ferns.

Conifers also occurred in the later Coal Measures, but more advanced plants – the flowering plants – first occur in rocks of Mesozoic age. Flowering plants produce seeds, and this group includes most of the familiar living plants such as palms, grasses, oak, ash, elm and all the vegetables. There are two main groups of flowering plants, which are distinguished by features of their seeds. Cotyledons are thick fleshy 'leaves' that form seeds. In monocotyledonous plants there is only a single seed 'leaf' in each seed. Maize, wheat and the other cereals are typical monocotyledons. The leaves of fossil and recent monocotyledons can be identified because the veins run along the leaves and are parallel to each other. Grasses and the leaves of *Nipa* from Sheppey (p. 81) show this clearly.

The other group of flowering plants is called the dicotyledons. These plants have two seed 'leaves' in their seeds. Beans and peas are typical of this group. Take the skin off a bean and you will see the two seed 'leaves' clearly. The leaves of dicotyledons have radiating veins that spread and form a network. The leaves of oak or a cabbage leaf show this very clearly.

Fossil wood, leaves and fruit all occur at Sheppey (p. 81), and we collected leaves from the coal measures in the Forest of Dean (p. 134).

Primitive plants include the seaweeds and many pond slimes, which are grouped as 'algae'. The earliest plants were algae, and their remains are known from rocks more than 3,000 million years old. We found fossilized algal growths in the old quarries at Portland (p. 107).

The word stromatolite means 'layer stone', which refers to the strange layered appearance that can be seen when these fossils are broken. Stromatolites are thought to be produced by algae which form deposits of calcium carbonate. We collected stromatolites from Perryfield Quarry on Portland (p. 106).

Algal growths at Old Quarries, Portland, Dorset

Stromatolite from Perryfield Quarry, Portland, Dorset

3 *Preparing to Collect*

EQUIPMENT

Most of the things you will need for collecting fossils are very simple, and you will probably already own many of them. At first it may be as well to visit a few sites using only the equipment that you already have, and then if you find that you enjoy collecting it will be worth spending a few pounds on the special tools and clothes that can make your collecting easier and more comfortable.

You should always dress properly for collecting. The British climate is notoriously unpredictable, and especially near the sea a sunny day can quickly turn to rain or become cold and overcast. In general, wear old clothes, because you will almost certainly get dirty. Walking or hiking boots are ideal, but wellington boots may be better in marshy or clayey areas. Wear jeans, a shirt with pockets and in the winter an old sweater and anorak. In summer a plastic raincoat is light and packs small enough to be worth carrying.

To carry extra clothes, equipment and fossils you will need a rucksack. Cheap canvas bags can be found at any army surplus store. Choose one that is comfortable and has a selection of side pockets. At sites such as Charmouth or Folkestone you can collect without any special equipment, but at other sites collecting is difficult or impossible without some tools. In the description of our collecting any special equipment needed is listed before each site is described.

Fossils found lying on the beach need only to be picked up, but those in large blocks of rock or in cliffs need careful removal. A hammer is needed at most sites and a geological hammer is best. It has a head with a square end on one side

The equipment you will need for serious collecting

and a chisel point on the other. A carpenter's hammer can be used, but the steel is usually too soft. A hammer weighing about a kilogram (2·2 lb.) is usually sufficient. Steel-shafted hammers are more expensive than those with wooden handles, but they are far less likely to snap when you are out collecting. As you gain experience you will find that a selection of cold chisels will be useful for cutting fossils out of the rocks; and a bolster – a broad-bladed chisel – is very useful for splitting rocks along bedding planes. At sites where the rock is soft a flat-bladed trowel or trenching tool may be useful. Finally, among the most essential items are wrapping paper, especially newspaper and tissues, kitchen foil, which is good for wrapping delicate specimens, cotton wool, sticky tape, polythene bags and an assortment of small boxes. If you are collecting seriously, then you will need a notebook to record details of finds and a felt-tip pen or magic marker to number specimens as you collect them.

A compass, tape measure and a variety of sieves may be useful for more advanced work. Maps are important to find sites, and geological maps may help you to find the age of the rocks in which you are collecting. A camera is always useful as a pictorial record of sites and good specimens will help you with your collecting.

FINDING COLLECTING SITES

Collectors in Britain are fortunate because in this country there is an almost complete sequence of rocks from the Cambrian onwards, and fossils can be collected from many of these. The most obvious sites are seaside cliffs, but there are also many inland quarries that are worth investigating. Small road cuttings and even temporary excavations for building should be visited in your search for fossils. In many areas, especially where the underlying rocks are of Mesozoic or Tertiary age, a collection of the commoner fossils can be built up very quickly.

The first step when collecting in a new area is to visit the local museum, which may have a collection of fossils. This will give you an idea of the range of local fossils that you can expect to find. It is also useful to see specimens, since this will help you to recognize those that are only partly exposed when you are collecting. The museum curator will often be able to advise you on the best places to visit. He may also put you in touch with local collectors or suggest local guide books.

Unfortunately there are many areas without a local museum exhibiting a fossil collection and without a geological guide book. In these areas you are thrown back on to your own resources and you will have to visit as many exposures as possible.

A list of the more easily available guides is given on p. 158.

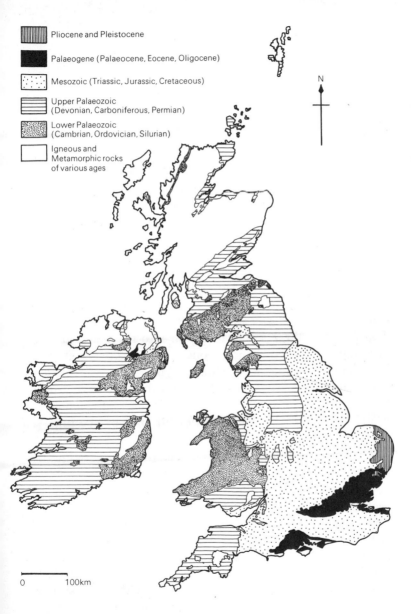

Pliocene and Pleistocene

Palaeogene (Palaeocene, Eocene, Oligocene)

Mesozoic (Triassic, Jurassic, Cretaceous)

Upper Palaeozoic
(Devonian, Carboniferous, Permian)

Lower Palaeozoic
(Cambrian, Ordovician, Silurian)

Igneous and
Metamorphic rocks
of various ages

0   100km

A geological map of the British Isles

HOW TO USE GEOLOGICAL SECTIONS

In the collecting diary we have included sections of the localities with the descriptions of collecting at each site. These should help you to find your way around the different beds of rock that you will see.

Cliff sections show the arrangement of the different beds of rock in relation to the bays and valleys along the cliffs. On these sections the vertical scale is greatly exaggerated in relation to the horizontal scale. If you use these sections with the Ordnance Survey map of the region you will be able to identify the different beds.

Let us look at the first of these sections (p. 73): if you walked westwards from Hackendown Point to Palm Bay you would cover a distance of over 2·5 km. Looking at the cliff at Hackendown Point you would be able to see three main beds. The lowest of these would disappear into the beach just west of Kingsgate and the middle one would disappear at White Ness, so that looking at the cliffs at White Ness you would see only a single bed, which would be the same as the upper bed at Hackendown. However, just west of White Ness a new bed would become visible high on the cliff, and by the time you reached Foreness this bed would make up most of the cliff.

The vertical sections that accompany these cliff sections give details of the beds as if they were piled one on top of the other in sequence. You will find these vertical sections particularly useful for finding beds that contain fossils. In the section on p. 73 the top bed is the one which appears at the top of the cliffs at White Ness and forms most of the cliff at White Ness. The bed below (3) disappears into the beach at Kingsgate, and the bed between (2) and (3) disappears into the beach at White Ness. The notes with the vertical sections will help you to find the beds that contain fossils. On the vertical section there is a bed with fossils indicated at (2). This is a 'Band full of echinoids, brachiopods and belemnites immediately above' it. Suppose that you wish to locate these bands when collecting. If you refer back to the cliff section you will see that the beds around (2) are exposed from east of Hackendown Point to White Ness, but at Hackendown Point they are high on the cliff and you are unlikely to be able to reach them; so to find this bed you should look just east of White Ness, where the bed will be near beach level.

In some cases (e.g. on p. 77) the technical names of the beds are given to the far right of the vertical section. These names are sometimes used in the text and faunal lists, and are widely used in other geological guides to the areas.

COLLECTING AT THE SITE

It is impossible to deal with all the problems that may be encountered while collecting, as each site presents its own special difficulties. There are, however, a few general methods of collecting that apply to all sites.

If the site is a working quarry or if you need to enter private land to reach a site, always ask permission. Where the quarry is owned by a large company you may find that you will have to write in advance for permission to collect. If you are crossing agricultural land, remember always to close gates behind you, walk near the edges of fields, and leave livestock alone. It is most important that you follow this advice, because thoughtless behaviour can prevent future access for other collectors.

When you reach the site try to identify the rock sequences exposed, using diagrams in guides or any other information that you have been able to obtain. Then examine the whole exposure quickly before looking at any part in detail. The debris at the bottom of the section will often reveal the presence of fossils; in many cases this debris also contains the best fossils. Fossils do not occur uniformly in sedimentary rocks: they may be scattered throughout the rock or they may be concentrated into particular beds. The preliminary examination of the site should reveal the situation that you have to deal with and may indicate the best spots or beds on which to concentrate your collecting. Some rocks yield fossils only after careful searching during many visits, so do not be disappointed if you do not find anything immediately. Searching for fossils requires patience, keen eyes and sometimes a little luck.

Once you have located the productive beds you can begin collecting in earnest. Part of the art of collecting lies in the correct use of your geological hammer, and the technique may have to be varied with each rock type that you meet. You will learn by experience how to tackle specific problems, but there are a few hints that should help you at the start. Wherever possible, try to split the rock along the bedding planes, since it will be on these that fossils are most likely to lie. To do this use either carefully placed blows from the square end of your geological hammer or split the rock using chisels. In more massive rocks, where the bedding planes are widely spaced, use the square end of the hammer alone or with chisels to split off small blocks. Since a fossil is a weakness in the rock cracks will tend towards and around it.

It is always worth cracking open any concretions that may be present, as these may contain fossils retaining their original shape while only crushed specimens occur in the surrounding rocks. For example, the ammonites in nodules from Port Mulgrave (p. 144) have their natural shape, but those in the shale are flattened and crushed. Different types of rock have different breaking characteristics, so it is often useful to practise a little at each new site to get the 'feel' of the rock.

You will find that the fossils can be extracted cleanly from some rocks with a few careful or lucky blows of your hammer; but usually only part of the fossil will be visible and the rest will be embedded in the rock. Do not try to extract fossils completely in the field. It is much safer to trim the surrounding rock and complete the extraction when you get home.

Collecting at Atherfield. Note the tools – hammer, chisel, scraper, pin vice and brush

Trimming can be done with the chisel end of the hammer or by using chisels. A bolster is particularly useful for trimming if the rock comes apart in thin slabs. If you break a fossil while extracting it, wrap the pieces together and if possible number the fragments and make a sketch of the original arrangement of the pieces. This will help you to reassemble the complete specimen later.

When you are collecting in soft sands or clays you are likely to find very fragile fossils, like the bivalve shells from Herne Bay (p. 78). These fossils must often be hardened before you collect them and a variety of materials can be used. If the specimens are dry, use Rawlplug Durofix diluted with equal parts of amyl acetate and acetone. This mixture is highly inflammable and gives a glossy surface to the specimens. If the specimens are wet, use a PVA emulsion glue. Evostik 'Woodworking Adhesive W' is useful and can be used dissolved in water. This glue is not suitable for use on very dry specimens because it forms a skin on the surface and does not penetrate as well as a glue that is dissolved in acetone.

Hardening materials should be applied carefully to fragile specimens. Very dilute glue may be dripped on to the fossils so that it is soaked up. Alternatively, use a small paint brush to spread glue on the fossil and surrounding sediment until the fossil is fully impregnated. Leave for 15–30 minutes to dry before extracting and wrapping to take home.

You will often find fragile specimens when you are not carrying the necessary hardening materials. In this situation, remove the specimens together with large blocks of the surrounding rock and clean them up at home. You can always take samples of unconsolidated rocks for sieving out small specimens.

Once you have completed the preliminary extraction of your specimen, the next step is to wrap it. Wrapping protects the specimens while you are carrying them home, and should a specimen break in transit the wrapping will keep the bits together. Newspaper is very good and cheap for this purpose, but specimens that are fragile should be wrapped in paper tissues or kitchen foil. If you find yourself without any of these, fragile specimens can be wrapped in anything soft, such as moss or dry sand. For greater protection the most fragile specimens should be placed in boxes or small tins. It is important to include labels with each specimen. These should give details of the locality where it was collected and if possible the level in the rock where it was found. Alternatively each specimen can be numbered, using a felt-tip pen, and all the important details can then be recorded in your notebook next to the specimen number. This may seem a waste of time at first, but it is worth taking a little trouble with labelling specimens; when your collection becomes large you may need to know where each specimen comes from so that you can identify weaknesses or strong parts of your collection. Specimens without their records are scientifically valueless.

Before moving to a new site tidy up the exposure by removing any discarded rock which could damage vehicles and farm machinery or injure livestock. If you leave dangerous rocks behind, you could be responsible for preventing access for other collectors. Do not try to strip a site of all its fossils, but be content with a few examples of each species. In many cases you should be able to gather a good representative collection from the debris at the foot of the exposure, and in that case you will not have to hammer the actual rock face. These last points are important because in some areas the thoughtless behaviour of a few collectors has destroyed valuable sites or has been responsible for the total banning of collecting.

4   A Collecting Diary

DAY 1   THE COAST OF KENT

To begin our collecting we visited four famous sites in Kent.
The day was showery with bright periods. While we were in
Folkestone it improved, becoming bright and sunny.

Botany Bay, Margate   Grid ref. T R. 392712

Botany Bay is about four kilometres east of Margate. We
travelled through the town and continued along the B2052,

Chalk cliffs at Botany Bay

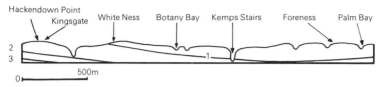

Hackendown Point · Kingsgate · White Ness · Botany Bay · Kemps Stairs · Foreness · Palm Bay

A cliff section of the Thanet coast between Kingsgate and Palm Bay

turning left into Percy Avenue, and at the end of this parking on Marine Drive. We had timed our visit so that we arrived on a falling tide, which allowed the longest time for collecting. We followed the narrow footpath down to the bay. There is a snack bar on the beach, which is useful if you intend to spend a long time collecting at this site. Botany Bay is pleasant to visit on a good day, but it was very cold and windy during our visit.

We turned right (east) at the end of the footpath and walked between the rocks into the first bay. The Chalk cliffs are

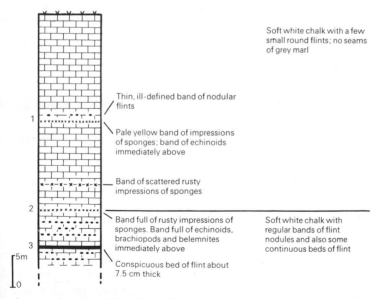

Soft white chalk with a few small round flints; no seams of grey marl

Thin, ill-defined band of nodular flints

Pale yellow band of impressions of sponges; band of echinoids immediately above

Band of scattered rusty impressions of sponges

Band full of rusty impressions of sponges. Band full of echinoids, brachiopods and belemnites immediately above

Conspicuous bed of flint about 7.5 cm thick

Soft white chalk with regular bands of flint nodules and also some continuous beds of flint

A vertical section of the Upper Chalk between Kingsgate and Palm Bay

weathered by the sea, and the lower part is blackened and encrusted with seaweed. We saw a few fossils in the bare patches in the lower levels, and we also found fossils in the Chalk pebbles on the beach. These fossils included molluscs and echinoids. The material of the fossils is harder than the Chalk, so that after only a little weathering they project from the rounded surface of the pebbles. Many of the pebbles were covered with holes. These are not fossils but are recent occurrences. Large holes about 1 cm in diameter are made by piddocks (*Pholas*), which are a kind of bivalve mollusc that bores into wood and rocks. Small holes about 1–3 mm across are made by *Cliona*, which is a sponge (p. 36) that bores into rocks.

The Chalk pebbles are very soft and can be easily broken or cut using a knife. This makes it easy to remove and prepare fossils from this site.

The best fossils occur in the higher parts of the cliff, where the Chalk has been weathered by the wind. We found them projecting from the Chalk and used hammer and chisels to remove them with a small amount of the surrounding Chalk.

Returning to the footpath we prospected to the west of the bay, where we located a bed containing brownish sponges, and about 1·5 m above this we found a bed that contained scattered flints. Between these two layers are rocks known as the *Echinocorys* beds, and here we saw crushed sea urchins and also collected a complete specimen of *Echinocorys* (p. 56). This bed can be traced through to Foreness Point and Palm Bay. The whole region of Chalk from White Ness Point (Grid ref. TR. 396710) to Foreness Point (Grid ref. 384717) is very productive, though it contains very few flints, which is unusual for Chalk of this age in southern England.

Commoner fossils from Botany Bay

A large fauna of fossils is known from the Chalk of this region. These fossils are of Late Cretaceous age and are therefore about 80 million years old.

Coral:	*Parasmilia, Caryophyllia, Epiphaxum, Axogaster*
Bryozoa	
Sponges:	*Siphonia, Cliona, Ventriculites, Plocoscyphia, Porosphaera, Doryderma*, etc.
Molluscs:	*Inoceramus, Spondylus, Pycnodonta, Lopha, Atreta, Plagiostoma, Chlamys, Neithea*
Brachiopods:	*Ancistrocrania, Kingena, Terebratula, Orbirhynchia*
Echinoderms:	crinoids, *Echinocorys, Micraster, Cyphosoma, Cidaris* (spines)

Notes on the fossils

The sponges occur as brown-stained patterned regions on the chalk. They are usually more or less circular and may be slightly raised above the surrounding rock. As the material of the fossil sponges is so diffuse they are difficult to collect, and the best that you can hope for is to get a fragment or to collect a sponge enclosed in a large block of chalk. We saw and photographed several sponges but did not try to collect any.

A sponge embedded in the chalk, Botany Bay

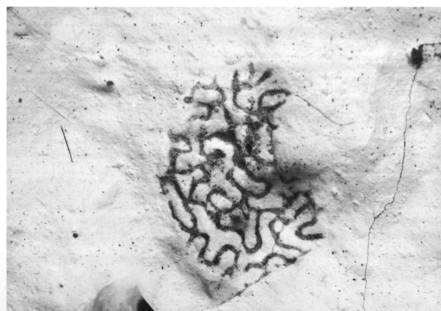

Inoceramus is a bivalve mollusc. Different species range in size from small shells like living clams to huge plate-like shells up to 1 m across. At Botany Bay complete shells of *Inoceramus* are not found, but fragments are very common. These are flattened pieces of greyish material that are harder than the chalk. One surface is smooth and the other is ridged. The broken edges show many fine ridges and grooves running through the full thickness of the fragment. These *Inoceramus* fragments may be up to 5 mm thick, but are rarely larger than 5 cm across.

Fossils from the Chalk can be cleaned using a needle or a penknife. Be careful when working near the surface of the specimen, as the fossils themselves are often very fragile. Echinoids can be easily cleaned with a needle, but remove the last fragments of chalk with a stiff toothbrush. Be especially careful when removing echinoids from blocks of chalk as they split easily.

Specimens from the chalk contain a lot of salt. Over about twelve months this will crystallize in the specimens and may cause them to break up. To avoid this, chalk fossils should be soaked in tap water for two or three weeks, with regular changes of water. This dissolves the salt out of the specimens, and after this simple treatment the fossils will last for a long time.

Herne Bay and Bishopstone Glen   Grid ref. T R. 212687

A small trowel will be useful at this site.

Travelling west from Margate along the A299, we turned right at Hawthorn Corner and left in Hillborough; right into Bishopstone Lane and left into Haven Drive. We parked near this turning and followed the public footpath to the beach. This footpath runs down a series of steps into a small valley known as Bishopstone Glen. We arrived on a rising tide, as we had previously visited Botany Bay, but you should try to arrive

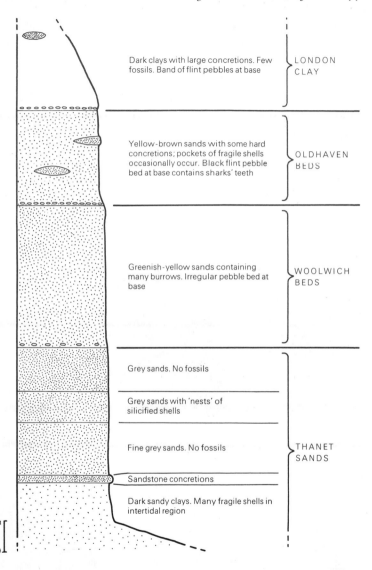

Dark clays with large concretions. Few fossils. Band of flint pebbles at base — LONDON CLAY

Yellow-brown sands with some hard concretions; pockets of fragile shells occasionally occur. Black flint pebble bed at base contains sharks' teeth — OLDHAVEN BEDS

Greenish-yellow sands containing many burrows. Irregular pebble bed at base — WOOLWICH BEDS

Grey sands. No fossils

Grey sands with 'nests' of silicified shells

Fine grey sands. No fossils

Sandstone concretions

Dark sandy clays. Many fragile shells in intertidal region — THANET SANDS

1m

0

A vertical section of the exposure at Bishopstone Glen, Herne Bay

when the tide is falling, because the most easily collected fossils occur low down on the beach.

At the bottom of the steps we turned right and walked down to the lower part of the beach, which is flat with extensive slightly raised areas of dark grey sandy clay. In this clay we found many shells of bivalve molluscs. These were whitish grey and they look like Recent clam shells. However, they have been leached (p. 14) and are therefore softer and more fragile than Recent shells. As a result they are difficult to collect. Specimens were collected together with large blocks of the surrounding clay, which helped to protect the shells until we

Collecting clay containing fossils on the beach at Herne Bay

got them home. The attached clay may later be sieved (p. 155) to extract smaller specimens. Large shells should be dried and hardened with a plastic glue before being washed out of the surrounding clay. They should then be soaked carefully to remove the salt. Although these shells are very modern in appearance they are Palaeocene in age and are thus about 65 million years old.

On the beach we also found harder bivalves. These had been washed out of clays that are blacker and occur further up the beach. Sharks' teeth were collected from small pools near the junction between the shingle and the clay (p. 30). Fossil wood also occurs in the clay, but it is very soft and we did not attempt to collect any.

When the tide had covered these clays we returned to Bishopstone Glen. This small gorge has steep sides of soft sandstone. About 3 m up the walls of the Glen there is a greyish sand. We dug along the surface of this and found groups or 'nests' of small bivalve molluscs. These molluscs are less than 1 cm across, but nests of them may be 10–15 cm across. The bivalves are reddish grey and are very hard, as they are silicified. We collected several nests of these molluscs in cloth bags and washed them from the surrounding sandstone when we got home.

Higher up in the banks of the Glen we saw a bed crowded with trace fossils. Most of these consisted of reddish-brown columns of sandy clay running through the grey sand. On closer examination we saw that these burrows were lined with a continuous cylinder of mud pellets. This kind of burrow is called *Ophiomorpha*, and similar modern burrow systems are made by prawn-like crustaceans. A second and less abundant kind of burrow is present. This consists of a simple infill of mud. We did not collect these trace fossils but we cleaned off the loose sand and mud with a soft brush and were then able to photograph them.

Just west of Bishopstone Glen and high up on the cliffs we found a band of black pebbles in brown sand. This band contains sharks' teeth; the teeth that we had found on the beach were probably from this bed. Sharks' teeth can be extracted from this bed by reverse sieving. To do this, a wide-meshed sieve, with holes from 1–2 cm in diameter, is used. Sediments from the pebble bed are washed through the sieve in water and the portion that passes through the sieve is then dried. Specimens may be found in these fine sediments.

Commoner fossils from Herne Bay

Molluscs: *Arctica, Corbula*
Vertebrates: *Odontaspis*

Notes on the fossils

Arctica is one of a group of bivalve molluscs that look like living clams. All have smooth shells with weakly defined growth lines as their strongest external feature.

   Corbula is a small bivalve – less than 1 cm long – which occurs in Bishopstone Glen (p. 44). Its shell has a long extension at one end.

   The teeth of the shark *Odontaspis* are long and slender with two roots. Well-preserved specimens have two small points near the base of the main centre point.

The Isle of Sheppey Grid ref. TR. 019726

Leaving Herne Bay, we continued west and then turned north on to the Isle of Sheppey. We took the A250 to Eastchurch, and by the church in the village we turned left on to Warden Road. We followed this road through to Warden Point. There is a café near the Point. We parked at the top of the cliff.

   There is very bad coastal erosion at Warden Point, and every year more and more land slips down to the sea, taking with it houses and wartime coastal defences. In winter and after long rainy periods the cliffs are extremely dangerous, because the clay is very soft. You must also keep careful watch on the tide, as it can easily cut you off. Warden Point is best visited at low tide, especially after frosty or long dry periods. Frost or very dry weather break up the clay, which falls into the sea. The hard fossils are then washed out and can be collected from the beach, where they are washed up and sorted by the action of the waves. Fossils can be found on the cliffs at Sheppey, but they are widely dispersed and difficult to see.

On the beach masses of brown pyritized fossils occur in shallow depressions and around large rocks. Most of these fossils are less than 5 cm across. Collecting is very easy. We looked through the concentrates of fossils and picked out good specimens; but you could collect bags full of the concentrate to be sorted at home. For collecting very small seeds you must sort the concentrate under a lens. Also on the beach we found large nodules which had trace fossils on their surfaces. These were mainly burrows similar to those at Herne Bay (p. 79), but here they were hard and could have been collected. Also on the beach and on the clay surface of the cliffs we saw large clear crystals of the mineral selenite.

Commoner fossils from Sheppey

Plants: fossil wood, leaves, fruit and seeds
Brachiopods: *Terebratulina*
Molluscs: many bivalves and gastropods
Vertebrates: *Lamna*, *Odontaspis*, *Striatolamia*

Notes on the fossils

Fossil wood is very common and occurs in pieces of all sizes. It is easily recognized as it looks like modern wood, and some pieces have knots or twigs attached. Fossil fruits are also very abundant, and during only a short period of collecting we found seven fruit of the palm *Nipa*, plus several other small

fruits or seeds. *Nipa* fruits are up to 5 cm long; they are
flattened, with one end pointed and the other rounded. Small
ridges run along the length of the fruit and often the surface
carries stronger folds as well. Pieces of *Nipa* fruit are also very
common and they often consist only of the blunt ends, which
are hollow.

   Terebratulina is a small brachiopod and the largest specimen

Nipa *fruit from the Isle of Sheppey*

that we found was only 1 cm long; most were only 5 mm long. The surface of this brachiopod carries many fine ridges, and the beak is clearly defined, with a large foramen on the beak of the pedicle valve. The two valves are equally convex but many specimens are crushed and distorted.

We found three kinds of sharks' teeth. The teeth of *Lamna* are large, with wide flattened crowns that have sharp edges sloping towards a sharp point. The two edges slope equally. The teeth of *Odontaspis* and *Striatolamia* are very similar. In both, the main part of the crown is long and slender with sharp edges and one face flattened, while the other is rounded. On each side of the main crown there is a small point, but these are often not preserved. Fish vertebrae were also collected; these look like the vertebrae of living fishes. Teeth of rays may be collected at Sheppey but we failed to find any.

Fossil wood borings also occur. These look like groups of small tubes and they are often hollow. These trace fossils represent the borings of the small bivalved mollusc *Teredina*, which is closely related to the living ship 'worm'.

Fossils from Sheppey need very little treatment or preparation. They should be dried thoroughly and then washed. This drying and washing breaks down and removes any clay that may still be attached to the specimens. Bits of pyrites cannot normally be removed without damaging the specimen, so it is better to be satisfied with what is visible.

Unfortunately pyritized fossils from Sheppey are liable to develop a condition known as pyrite disease (or 'pyritization'), which may very quickly break down the specimen to a fine dust. It is caused by the chemical decomposition of the pyrites in damp conditions; bacteria may play a part in this decomposition. Water in the atmosphere slowly reacts with the pyrites to form sulphuric acid. This then reacts with more pyrite and forms iron sulphate.

Pyrite disease does not, however, occur in all specimens from Sheppey, which is fortunate as there is little that the amateur collector can do to treat this condition when it occurs.

Folkestone

We now crossed south-east England to Folkestone, where we parked on the cliffs to the east of the harbour. We walked east along the beach until we were about a kilometre from the harbour. Here we saw some large concrete posts standing on the beach. These are the remains of an old sewage outlet and the best collecting begins to the east of them.

Climbing up the cliffs we found the junction where the dark grey Gault clay meets the yellowish Lower Greensand below. This junction is defined by a band of brown or black stones. The junction is high up the cliffs at the harbour end but it gradually becomes lower and lower until near the sewage outlet it is only about 3 m above the beach. Most of the fossils on the beach have been washed out of the Gault clay but we found one echinoid – *Hemiaster* – that came from the Lower Greensand.

Fossils in the cliff can be found by simply cracking open dry pieces of the Gault clay. These fossils usually have a lustre like mother of pearl. They are soft and difficult to collect, and you should always remove a block of clay around any good specimens. If you have a long time for collecting then spend an hour or so looking on the cliffs, because you may find some fine ammonites. It is easiest to collect on the cliffs after long periods of wet weather; but the clay will then be very soft so you will need old clothes. When we were collecting, the clay was dry, but we found several good bivalve molluscs and some ammonites.

The quickest collecting is on the beach, but this area cannot be reached at high tide. As at Sheppey, the sea washes the fossils out of the clay and concentrates them in depressions around boulders on the beach. Most of the fossils are hard. Those higher up the beach have been least worn by the sea and are therefore usually better than those lower down. Collecting

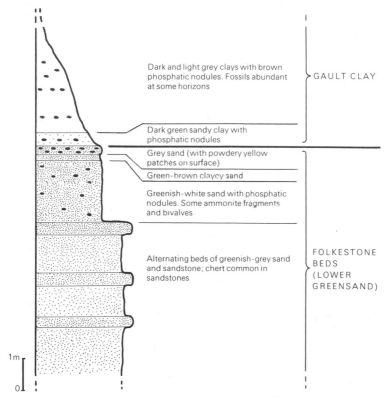

Dark and light grey clays with brown phosphatic nodules. Fossils abundant at some horizons — GAULT CLAY

Dark green sandy clay with phosphatic nodules

Grey sand (with powdery yellow patches on surface)

Green-brown clayey sand

Greenish-white sand with phosphatic nodules. Some ammonite fragments and bivalves

Alternating beds of greenish-grey sand and sandstone; chert common in sandstones — FOLKESTONE BEDS (LOWER GREENSAND)

1m

0

A vertical section of the Gault–Lower Greensand junction at Copt Point, Folkestone

simply involves picking through the concentrations of fossils in the rock pools, and as the fossils are hard they only need wrapping. They should be soaked when you get them home to remove the salt.

Commoner fossils from Folkestone

This list is divided into fossils occurring in the Lower Greensand and those from the Gault Clay. However, the fossils on the beach represent a mixture from these two levels.

LOWER GREENSAND

Molluscs:	bivalves:	*Inoceramus, Panopea, Nanonavis, Cucullaea, Thetironia, Resatrix, Pseudocardia, Pterotrigonia, Linotrigonia, Entolium, Neithea, Exogyra, Gryphaeostrea*
	gastropods:	*Anchura, Tessarolax, Eucyclus, Metacerithium, Mesalia, Leptomaria, Gyrodes*
	cephalopods:	nautiloids: *Eutrephoceras* ammonites: *Douvilleiceras, Beudanticeras*

Notes on the fossils

A single piece of fossil wood was collected during our visit. This wood had a grain and texture like modern wood, but more detailed identification was not possible.

Molluscs are very common at Folkestone. Four species of *Inoceramus* are known, but we found only two. *Inoceramus sulcatus* ranges from 1 to 5 cm long. It has a pointed beak, and its shell widens away from the beak. However, the shell is more deep than long and often twists so that the beak is flexed towards the front. The specific name *sulcatus* refers to the very strong ridges or sulcae that run along the shell and are characteristic of this species. *Inoceramus concentricus* has a smoother shell that carries many concentric ridges or growth lines. This species is common and again ranges from 1 to 5 cm long. The beak curves over the short hinge and in good specimens with both beaks preserved the beaks almost meet.

Gastropod molluscs are fairly common. Most of the specimens found on the beach consist of internal moulds, but those from the cliffs often have the shell preserved. *Anchura*

GAULT CLAY (*this list shows the commoner forms only and does not include any microfossils*)

Annelids:		*Serpula, Hamulus, Rotularia, Sarcinella, Glomerula*
Molluscs:	bivalves:	*Inoceramus* (4 species), *Nucula* (3 species), *Acila, Messosacella, Entolium, Plicatula, Pseudocardia, Nanonavis, Linotrigonia, Corbula*
	gastropods:	*Nummocalcar, Leptomaria, Conotomaria, Pleurotomaria, Anchura, Tessarolax, Buccinofussus, Sipho, Gyrodes, Mesalia, Metacerithium, Confusiscala, Torquesia*
	scaphopods:	*Dentalium, Cadulus*
	cephalopods: ammonites:	*Beudanticeras, Hoplites, Anahoplites, Dimorphoplites, Callihoplites, Euhoplites, Prohysteroceras, Hysteroceras, Heteroclinus, Anisoceras, Idiohamites, Hamites*
	belemnites:	*Neohibolites*
Arthropods:	crustaceans:	*Notopocorystes*
Fishes:		*Lamna*

has a high spire, and very well-preserved specimens may show a long canal and a large flange from the outer lip of the aperture. However, the only specimens of *Anchura* that we collected were internal moulds.

Specimens of *Dentalium* (p. 50) are usually hollow, with a circular cross-section and fine ridges on the outer surface. They usually consist of almost straight cylinders.

Specimens of at least nine species of ammonites were collected. Complete ammonites are relatively rare on the beach, and only six of our seventy-five specimens showed complete coils. The other specimens were all short fragments. Ammonites collected from the cliffs had their shells preserved, and these had a lustre like mother of pearl. On the beach most of the ammonites consisted of internal moulds. The sutures were very clearly shown on some of the ammonites collected, while many specimens were broken along the septum and consisted only of isolated ammonite chambers.

A fragment of an ammonite (Beudanticeras) *showing suture line which are etched by weathering*

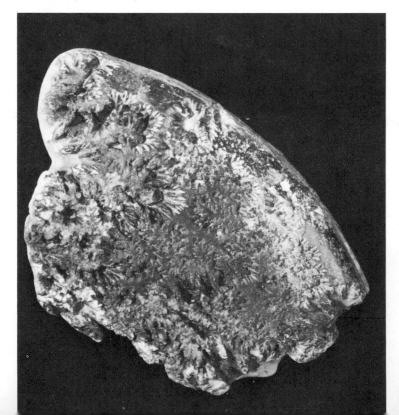

DAY 2 THE COAST OF DORSET

Our second day was devoted to collecting along the coast of Dorset. Several classic localities are found on this coast, as well as some of the richest collecting in Britain. We spent the night at Bridport and made a very early start, breakfasting at seven o'clock and arriving at our first locality by 8.15. As we did not finish collecting until seven in the evening we got in many hours at seven different sites, and we finished our day with a large collection of fossils from Jurassic and Pleistocene deposits. Most of the day was foggy with brief showers, but later the weather cleared and our visit to Portland was pleasant.

Charmouth   Grid ref. SY.364930

Take as many small boxes as you can get to this site.

We travelled west along the A35, and at the traffic lights in Charmouth town we turned left into Lower Sea Road. The beach is about a kilometre down this road and we parked near the sea. The river Char flows into the sea to the east of the car park. This river is usually shallow and can be forded easily if you are wearing walking boots. If the river is too deep there is a footbridge about 200 metres inland. After crossing the river we walked eastwards along the shore for about $1\frac{1}{2}$ kilometres. This took us twenty minutes.

We saw fossils all along the beach, and we saw many worn pebbles with bits of ammonites, gastropods and fragments of belemnites showing. These pebbles are usually light grey, while the fossils are white.

The cliffs consist of black-grey clay for the lower forty metres. This is the Black Ven Marl, and above it there is a grey marl that is known as the 'Belemnite Marl' on account of the

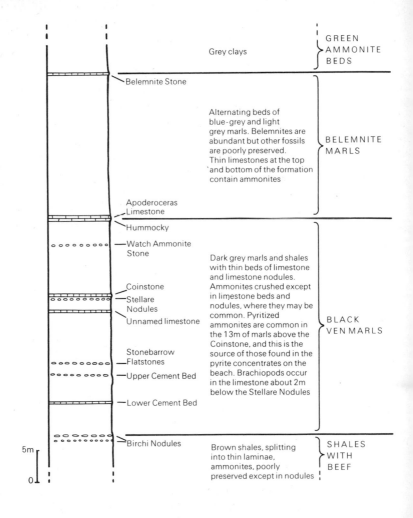

GREEN
AMMONITE
BEDS

Grey clays

Belemnite Stone

Alternating beds of
blue-grey and light
grey marls. Belemnites are
abundant but other fossils
are poorly preserved.
Thin limestones at the top
and bottom of the formation
contain ammonites

BELEMNITE
MARLS

Apoderoceras
Limestone

Hummocky

Watch Ammonite
Stone

Dark grey marls and shales
with thin beds of limestone
and limestone nodules.
Ammonites crushed except
in limestone beds and
nodules, where they may be
common. Pyritized
ammonites are common in
the 13m of marls above the
Coinstone, and this is the
source of those found in the
pyrite concentrates on the
beach. Brachiopods occur
in the limestone about 2m
below the Stellare Nodules

Coinstone

Stellare
Nodules

Unnamed limestone

BLACK
VEN MARLS

Stonebarrow
Flatstones

Upper Cement Bed

Lower Cement Bed

Birchi Nodules

Brown shales, splitting
into thin laminae,
ammonites, poorly
preserved except in nodules

SHALES
WITH
BEEF

5m

0

A vertical section of a point below Stonebarrow, near Charmouth

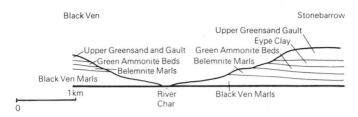

A cliff section of the Dorset coast between Black Ven and Stonebarrow

large number of belemnites that it contains. The cliffs at Charmouth are very dangerous; they are high and pieces of rock are continually falling off them. Large rock falls or landslips occur frequently. There are also many mud flows, which may be treacherous and are always very dirty. We did not attempt to collect on the cliffs but concentrated on the beach, where there is by far the easiest and richest collecting. Landslips bring large sections of the cliff below the high-tide level, so that twice each day the sea washes fossils out of the clays and marls. These fossils can be found as concentrations near the landslips and can be collected at any time except high tide.

This is one of the richest places in Britain to collect belemnites, but ammonites (p. 48) and pieces of crinoid stem (p. 57) are also common, and there is a large fauna of other fossils, including vertebrates. Large ammonites can be seen in the bigger boulders on the beach, but these are too large to collect.

Like the fossils from Sheppey (p. 81) many of the Charmouth fossils are made of pyrites. Larger pieces of pyrites may also be found on the beach, and some of these are worth collecting. There is a band of pyrites in the cliffs, but it is usually hard to locate. Fossils from Charmouth are subject to pyritization (p. 83), but this seems to be selective: some specimens may be affected quickly, while apparently similar specimens may survive for many years.

Commoner fossils from Charmouth

COMMONER FOSSILS FROM THE BLACK VEN MARLS

Brachiopods: *Piarhynchia, Tropiorhynchia*
Molluscs: gastropods: *Pleurotomaria*
               bivalves: *Plagiostoma, Oxytoma, Gryphaea*
               ammonites: *Asteroceras, Arnioceras, Caenisites,*
                               Crucilobiceras, Cymbites,
                               Epophioceras, Eoderoceras,
                               Epideroceras, Echioceras, Gleviceras,
                               Microderoceras, Oxynoticeras,
                               Promiceroceras, Xipheroceras
Echinoderms: crinoids: *Pentacrinites*

COMMONER FOSSILS FROM THE BELEMNITE MARLS

Molluscs: gastropods: *Amberleya*
               bivalves: *Parainoceramus, Chlamys*
               ammonites: *Apoderoceras, Tropidoceras,*
                               Phricodoceras
               belemnites: *Hastites, Pseudohastites,*
                               Passaloteuthis, Angeloteuthis
Brachiopods: *Cincta, Tropiorhynchia*
Trace fossils: *Chondrites*

    If you are staying in or near Charmouth, Barney's Shell Shop and Fossil Exhibition is well worth a visit. It is in Charmouth town beside the A35, about 100 metres down the hill from the traffic lights. It is well signed and has one or two large ammonites in the window. For only a few pence you can see many fine specimens from the Charmouth area that have been collected by Barney over the past few years.

    Lyme Regis Museum is also worth a visit. Here there are

many good specimens displayed, including ichthyosaurs and plesiosaurs. The beach to the east of Lyme Regis is a classic area from which Mary Anning made her famous and very important collections during the first half of the nineteenth century. Fossil collecting is, however, relatively poor there today.

Seatown Grid ref. SY.419916

We travelled east along the A35 and in the village of Chideock turned right, following the signs to Seatown. After about a kilometre we reached Seatown beach and parked by the road. On the beach we turned right and walked westwards to the point. This was a twenty minute walk, but we found fossils along the cliffs, including belemnites, oysters (*Inoceramus* and *Gryphaea*) and ammonites. At the point there is a wave-cut platform on which very rich beds of belemnites (p. 46) are exposed. This platform is cut into the Belemnite Marl already encountered at Charmouth. Here at least six different kinds of belemnites may be found, and at a single glance we were able to see literally hundreds of specimens.

Ammonites are found in the cliffs in what is called the 'Green Ammonite Beds'. It is usually safe to climb on the cliffs here to collect them. Looking up at the cliffs from the beach you will be able to see three bands of hard rock that

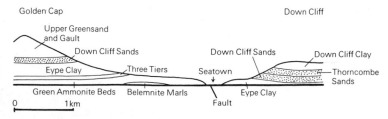

A cliff section of the Dorset coast between Golden Cap and Down Cliff

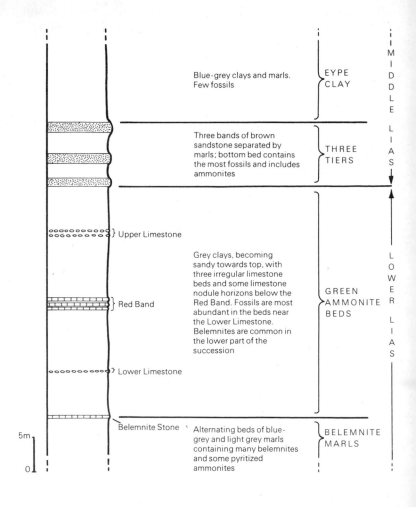

Blue-grey clays and marls. Few fossils — EYPE CLAY

Three bands of brown sandstone separated by marls; bottom bed contains the most fossils and includes ammonites — THREE TIERS

Upper Limestone

Grey clays, becoming sandy towards top, with three irregular limestone beds and some limestone nodule horizons below the Red Band. Fossils are most abundant in the beds near the Lower Limestone. Belemnites are common in the lower part of the succession — GREEN AMMONITE BEDS

Red Band

Lower Limestone

Belemnite Stone

Alternating beds of blue-grey and light grey marls containing many belemnites and some pyritized ammonites — BELEMNITE MARLS

MIDDLE LIAS

LOWER LIAS

5m
0

A vertical section of the Lower and Middle Lias between Seatown and Golden Cap

project from the cliff face. These are a useful marker, as they indicate the top of the Green Ammonite Beds. The bands of rock extend from St Gabriel's Mouth (Grid ref. S Y. 395922) to Seatown.

We did not try to collect many specimens at Seatown, since in general these would have duplicated our Charmouth collection. But the site is well worth visiting on a good day: collecting on the cliffs and in the Belemnite Marls on the beach can be pleasant and should produce a fine collection.

Eype   Grid ref. S Y. 447910

We continued east along the A35 and in Miles Cross, about $1\frac{1}{2}$ kilometres before Bridport, turned right, following the signposts to Eype. At the first T-junction we turned left to Eype Mouth and after about 100 metres took the first right. This was marked 'No through road', but it leads to Eype village, where we turned right, passing the Eype's Mouth Hotel on the right and continuing down to the beach. We parked on top of the cliffs and took the steps to the beach, where we turned right and walked westwards for about a kilometre to the point. Here we found blocks of yellow sandstone that had fallen to the beach. These blocks are from what is known as the Thorncombe Sands and in them we saw scallops, belemnites

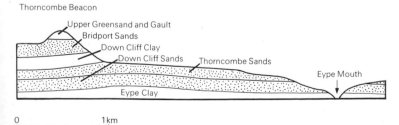

A cliff section of the Dorset coast between Thorncombe Beacon and Eype Mouth

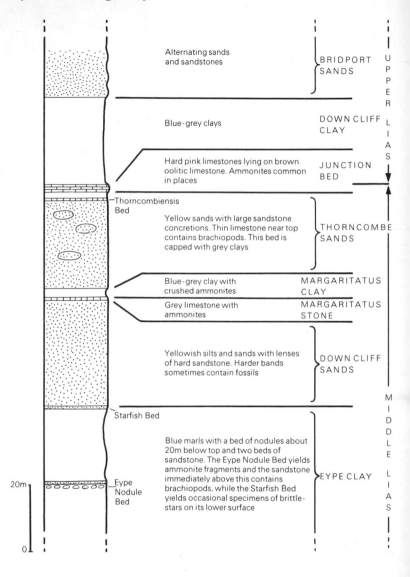

Alternating sands and sandstones — BRIDPORT SANDS

Blue-grey clays — DOWN CLIFF CLAY

Hard pink limestones lying on brown oolitic limestone. Ammonites common in places — JUNCTION BED

Thorncombiensis Bed

Yellow sands with large sandstone concretions. Thin limestone near top contains brachiopods. This bed is capped with grey clays — THORNCOMBE SANDS

Blue-grey clay with crushed ammonites — MARGARITATUS CLAY

Grey limestone with ammonites — MARGARITATUS STONE

Yellowish silts and sands with lenses of hard sandstone. Harder bands sometimes contain fossils — DOWN CLIFF SANDS

Starfish Bed

Blue marls with a bed of nodules about 20m below top and two beds of sandstone. The Eype Nodule Bed yields ammonite fragments and the sandstone immediately above this contains brachiopods, while the Starfish Bed yields occasional specimens of brittle-stars on its lower surface — EYPE CLAY

Eype Nodule Bed

UPPER LIAS

MIDDLE LIAS

20m

0

A vertical section of the Upper and Middle Lias sequence between Eype Mouth and Thorncombe Beacon

and *Pentacrinus*. There were also good trace fossils on some blocks. These were *Thalassinoides* – crustacean burrows (p. 15). We found a few specimens of the ammonite *Amaltheus* in the clays at the lower part of the cliffs.

Our visit to Eype was not very successful. Several guide-books suggest that this is a rich locality but we found it disap-pointing. It is a long way from the main road; fossils are difficult to collect from the Thorncombe Sands and relatively rare in the clays. If you are near Eype the fossil-bearing areas may be worth a visit but the site is not worth a special journey. However, the conditions for collecting do vary at Eype; it is dependent on rock falls to maintain a supply of new blocks on the beach. During our visit there was very little of the rich Junction Bed (see p. 96) available on the beach.

Commoner fossils from Eype

COMMONER FOSSILS FROM THE EYPE CLAY

Molluscs:	bivalves:	*Oxytoma, Chlamys, Lucina, Astarte,*
		Grammotodon, Lima, Parallelodon,
		Pleuromya, Protocardia, Pseudolimea
	gastropods:	*Procerithium, Pleurotomaria*
	ammonites:	*Tragophylloceras, Amaltheus,*
		Leptaleoceras, Amauroceras,
		Lytoceras, Liparoceras, Metacymbites
Coral:		*Montlivaltia*
Brachiopods:		*Lingula, Spiriferina, Furcirhynchia*
Echinoderms:	brittle star:	*Ophioderma*

COMMONER FOSSILS FROM THE DOWN CLIFF SANDS

Brachiopod:		*Gibbirhynchia*
Molluscs:	bivalves:	*Plicatula, Pseudopecten, Gryphaea*
	ammonites:	*Amaltheus, Lytoceras*
	nautiloids:	*Nautilus*
Echinoderms:	crinoids:	*Isocrinus*

COMMONER FOSSILS FROM THE MARGARITATUS CLAY
Molluscs: ammonites: *Amaltheus*
                 some crushed bivalves

COMMONER FOSSILS FROM THE THORNCOMBE SANDS
Brachiopods: *Gibbirhynchia*
Molluscs: bivalves: *Gryphaea, Pseudopecten*
               ammonites: *Amaltheus, Amauroceras*
               belemnites: *Belemnites*
Trace fossil: *Thalassinoides*

COMMONER FOSSILS FROM THE JUNCTION BED
Brachiopods
Gastropods
Molluscs: ammonites: *Pleuroceras, Dactylioceras,*
                            Hildoceras, Harpoceras,
                            Harpoceratoides

Burton Bradstock (Freshwater)   Grid ref. SY.477895

Take a heavy hammer and a chisel to this site.

   Continuing east on the B3157 from Eype to Burton
Bradstock, we passed a large camping site on the right-hand
side of the road and then turned right following the signpost to
Southover. After passing the Dove Inn we parked beside the
road and then followed the public footpath to Freshwater. At
the beach we turned left and walked eastwards. The cliffs are
made of yellow sandstone that does not contain fossils, and we
ignored blocks of this on the beach. However, about 200
metres along the beach we were able to see a band of hard dark
brown rocks that were visible high up on the cliffs. This band

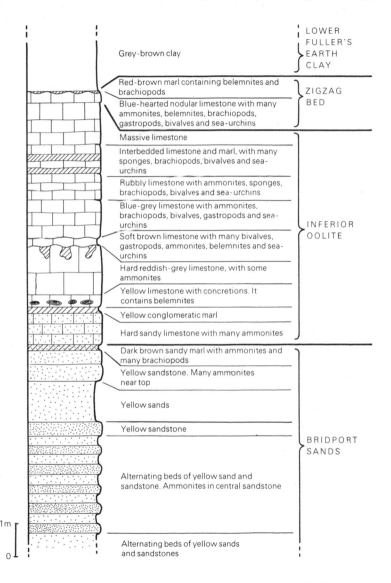

A vertical sequence of the Inferior Oolite sequence, Burton Cliff.

Cliffs at Burton Bradstock

continues for about a kilometre and is very rich in fossils. It cannot be reached, however, and must therefore be studied and collected from blocks that have fallen to the beach. We found a large fall of these blocks and were able to collect many good specimens.

It is difficult to extract fossils from the hard rock, but this is the richest area in Britain for collecting specimens of ammonites from the Inferior Oolite, which is about 175 million years old. As this is a classic site, you may like to simply see and photograph the blocks. But it is quite easy to collect small blocks of this rock and to break larger ones with your sledgehammer. We collected several small pieces of rock and were able to extract several specimens from them using a hammer and small chisel.

Portland Bill   Grid ref SY.677682

Leaving Burton Bradstock we took the B3157 east. In Weymouth we followed signs to Portland, and after crossing the bridge on to the island we continued towards Fortuneswell.

Commoner fossils from Burton Bradstock

Sponges:		*Peronidella, Leucospongia, Elasmostoma*
Brachiopods:		*Aulacothyris, Sphaeroidothyris, Homeorhynchia, Cincta*
Molluscs:	bivalves:	*Astarte, Lima, Pholadomya, Pleuromya, Variamussium, Anisocardia, Nuculana, Opis*
	gastropods:	*Pleurotomaria, Trochus*
	ammonites:	*Leioceras, Bredyia, Tmetoceras, Brasilia, Emileia, Stephanoceras, Strenoceras, Spiroceras, Garantiana, Dimorphinites, Polyplectites, Strigoceras, Oecotraustes, Oxycerites, Procerites, Planisphinctes, Parkinsonia, Polysphinctes, Ebrayiceras, Morphoceras, Zigzagiceras*
	nautiloids:	*Cenoceras*
	belemnites:	*Belemnites*
Echinoderms:	echinoids:	*Collyrites, Clypeus, Pygorhytis, Holectypus, Stomechinus*

The west side of the island is a very good collecting area, as many tons of rubble from the Portland Stone Quarries were dumped along the base of the cliffs and this rubble contains many fossils. We did not visit that area on this trip but continued to Easton. There is a museum in Easton that is worth a visit, since it contains good specimens of large ammonites and fossils from Portland. From Easton we continued to Southwell and turned left into the High Street. Portland Bill is about $2\frac{1}{2}$ kilometres along this road. We parked in the large car park and

A fallen block containing fossils at Burton Bradstock

Leioceras, *an ammonite, from the fallen block shown above*

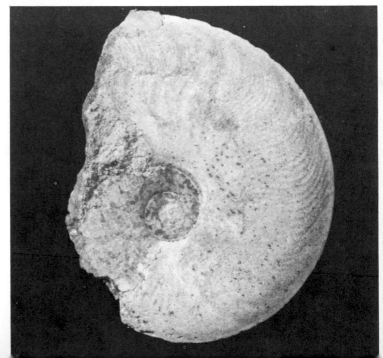

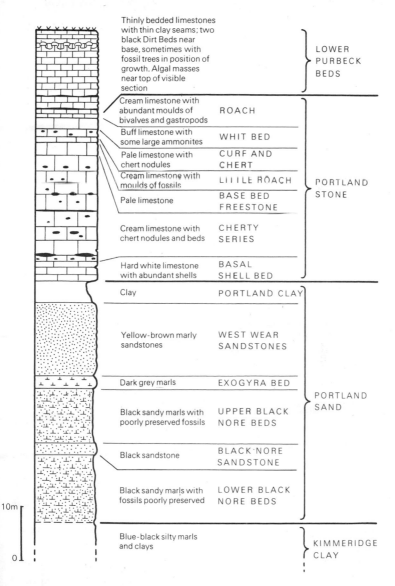

A vertical section through the Portland and Lower Purbeck Beds of the Isle of Portland

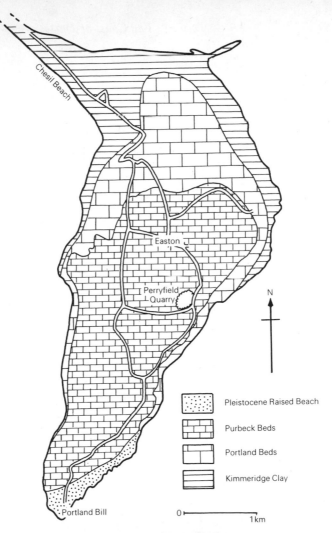

A geological sketch map of the Isle of Portland

walked across the grass towards the red-and-white-banded
lighthouse.

The whole end of Portland Bill is covered with Pleistocene
deposits that form a raised beach. These deposits are exposed
at the top two metres around the cliffs, but collecting is richest
to the east of the lighthouse and behind the Lobster Pot Café.

The fossils include many shells, particularly limpets and other gastropods. Since many of these look like modern shells, it is possible to confuse them with shells that have been carried up the cliffs. If, however, you are always careful to dig your specimens out of the soft sandstone that is packed with fragments and complete shells, then you can be sure that you are collecting only Pleistocene specimens.

We collected many shells from these deposits, and we also collected a bag of sand that we took home for sieving to extract small specimens. These raised beach deposits indicate a sea temperature slightly cooler than that found today in the Portland area; similar assemblages of molluscs are found today in northern England and Scotland.

We then walked along the cliffs away from the lighthouse. We passed a small crane over a deep area that was used for loading rock on to boats, and just past this we found fossil oysters in the rocks. These oysters occur in rocks at the top of the Portland Stone, which was laid down about 140 million years ago.

Portland Bill, showing the raised beach

Commoner fossils from Portland Bill

The commonest fossils in the raised beach deposits are:

Molluscs: gastropods: *Neritoides, Littorina, Rissoa, Patella*
bivalves: *Mytilus*

Perryfield Quarry   Grid ref. SY.690703

Leaving Portland Bill we travelled back towards the mainland
and stopped at Perryfield Quarry, which is a large stone quarry
opposite the entrance to Cove Chalet Park. If you intend to
visit this quarry it is wise to obtain permission from the quarry
manager a few days beforehand. We parked in the entrance to
the quarry and made our way around the quarry machinery to
the south end of the workings. Here there are many
stromatolites, which occur high up on the face of the workings;
they regularly fall into the quarry and may be picked up
around the first level of the workings. Stromatolites are oval or
rounded, but there are many broken fragments scattered
about. They range in size from 20 cm to 70 cm across, and
their upper surfaces are deeply pitted and roughened, with
rounded raised areas. Broken stromatolites show an obvious
layered structure. We collected several small stromatolites (p. 61).

Old Quarries, Portland   Grid ref. SY.690727

About two kilometres after Easton on the right-hand side of
the road there is a large car park, and on the face of the
building bordering this is a plaque that reads *Bath & Portland
Stone Firms Ltd*, while the building next to this carries the
trade sign *Vickers Ltd*. Behind these buildings is a 'fossil
garden', which contains fossil tree trunks and wood as well as
four large ammonites (*Titanites*) that are up to a metre across.

Ammonites and fossil tree trunks in the fossil garden at Portland

After looking round the garden we followed the footpath round the back of the building and into the disused quarry workings behind the fossil garden. We walked parallel to the road, less than fifty metres from it, and after about one hundred metres we saw on the right-hand side of the path several circular to oval structures that were from two to three metres across (p. 61). Each of these consists of a wide band of rock, about 0·3 m thick, surrounding a hollow region. These rings, which are raised about 0·3 m above the grassy surface, represent algal growths that developed around the bases of the trunks of cycad trees. Fossil wood has been found at the bottoms of the hollows. Structures similar to these may be seen in the fossil forest at Lulworth.

We then walked 100 or 200 metres further away from the road. Many of the rocks along the path held fossils. These rocks contain many small hollows, spirals of rock and fossil oysters, and they form a definite bed that is limited to only a part of the larger blocks. This bed is called the 'Roach'; it

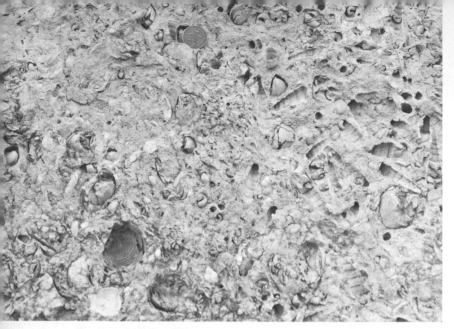

Fossils in 'the Roach'

marks the top of the Portland Stone and therefore corresponds
to the oyster bed that we saw at Portland Bill (p. 105). The bed
is extremely rich in fossils and is particularly interesting
because the fossils consist almost entirely of internal and external moulds from which the material of the animals has been
dissolved away to leave only hollows. In the Roach the bivalves
Ostrea and *Camptonectes* may be represented as casts, but the
bivalve *Trigonia* is usually represented only as moulds. The
external moulds of *Trigonia* have surfaces with many small
pin-prick indentations, while the internal moulds usually show
the hinge teeth (p. 44), which are characteristic as they consist
of a pair of large flanges (hollows in the moulds) that meet in a
'V'. The faces of these teeth carry fine vertical ridges. Internal
moulds of *Trigonia* were called 'osses heads' by Portland
quarrymen, presumably because they vaguely resemble the
head of a horse. Another common fossil is an internal mould of
the gastropod *Aptyxiella*. This consists of a long spiral of rock
representing an internal mould, and was known as the Portland
Screw.

We photographed some specimens of the Roach and collected some small blocks. We then finished collecting for the day.

Commoner fossils from the Roach
Molluscs: bivalves: *Laevitrigonia, Myophorella,*
                                  Camptonectes, Protocardia
            gastropods: *Aptyxiella*

DAY 3   THE ISLE OF WIGHT

The Isle of Wight is one of the best areas in Britain for collecting fossils, as it has a range of rocks of Mesozoic and Cainozoic age which contain many different kinds of fossils.

We started collecting at eight o'clock in the morning and finished at six. In this time we visited four main localities and one smaller site. Numbers of specimens collected are meaningless, but during the day we saw hundreds of thousands of specimens that could have been collected; in some places the rocks consist almost entirely of fossil molluscs.

Whitecliff Bay   Grid ref. SZ.640857 – SZ.645865

Take cloth and polythene bags with you to this site.

From Sandown we took the A3055 towards Brading and turned right at Yarbridge on to the B3395. After about four kilometres we turned right, following the signposts to Whitecliff Bay. We turned right at the duck pond and after three kilometres parked near the T-junction. From here we took the footpath to the cliff top and walked 400 metres along the top of the cliffs until we reached a footpath down to the beach. If you are facing the sea there are high cliffs to your right (south). We spent over an hour collecting at the base of

Whitecliff Bay

these cliffs, breaking the Chalk boulders that litter the beach and checking their sea-worn and broken surfaces for fossils.

Commoner fossils from the Upper Chalk at Whitecliff Bay

Sponges: *Porosphaera*, other unidentified sponges
Bryozoans
Molluscs: *Inoceramus, Pecten, Belemnitella mucronata*
Brachiopods: *Magas, Crania, Kingena*
Echinoderms: *Echinocorys, Cardiaster*, Cidarid echinoids

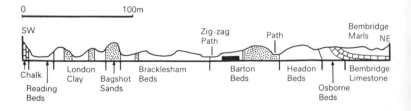

A cliff section of Whitecliff Bay

Description	Formation
Green clays with silts, with a pale sand at base which contains fossils	BEMBRIDGE MARLS
Cream limestones with moulds of freshwater and occasionally land gastropods	BEMBRIDGE LIMESTONE
Green clays with conspicuous beds of red clay near top. Few fossils	OSBORNE BEDS
Grey and green clays with thick sequence of brown and yellow sandy clays near the middle. Brown sandy clay about 8m above base contains abundant marine molluscs. Remainder contains scattered freshwater and brackish water molluscs	HEADON BEDS
Grey-green clays and sandy clays overlain by a thick sequence of yellow-brown sands. Upper group of sands does not contain fossils. Sandy clays and clays below are poorly exposed but fossils can be collected in scattered exposures immediately south of the metalled cliff path	BARTON BEDS
Thick sequence of grey and green clays and sandy clays with some yellow sands. Conspicuous band of grey clays with lignites occurs about 54m above base. Dark grey sandy clays extending for some 16m below this contain many molluscs. Green sandy clays above the lignitic horizon contain many molluscs, and about 20m above the clay is a band containing many large *Nummulites*. Sandy clays and clays between about 135 and 140m above the base contain molluscs and small *Nummulites*. Base marked by a conspicuous pebble bed	BRACKLESHAM BEDS
Yellow and brown sands	BAGSHOT SANDS
Brown and grey-brown clays with two bands of pale sands. Lowest 5m is a greenish sandy clay with worm tubes in lenses. Sharks' teeth and some molluscs are present. Fossils present but poorly preserved in remainder of sequence	LONDON CLAY
Red clays with a pale sand immediately above the chalk. No fossils	READING BEDS
Hard white limestone with sponges, echinoids and occasional belemnites	UPPER CHALK

100m

0

A simplified vertical section of the Tertiary succession exposed at Whitecliff Bay

Notes on the fossils

Sponges, sea urchins, *Inoceramus* and fish were collected. *Porosphaera* is a spherical sponge from 0·5 to 2·0 cm in diameter, and its surface carries very small 'pin-prick' pores. This is the only sponge that is readily identifiable from Whitecliff Bay. We collected several specimens of *Porosphaera* and also saw other sponges enclosed in flint nodules. If the flint nodules are broken open many of them are found to be hollow. The flint is black but the region around the hollow is white and heavily pitted. This is the fossil sponge.

Inoceramus is common. At Whitecliff Bay a small species of the genus occurs. It has a shell up to 5 cm long which carries many concentric ridges. A single species of belemnite – *Belemnitella mucronata* – occurs at Whitecliff Bay. This is a zone fossil and is used to correlate the deposits forming one part of the Upper Chalk in Great Britain.

We found the sea urchin *Echinocorys*, which we had previously collected from Botany Bay (p. 74); on its surface there were patches of a bryozoan.

Fossil fishes also occur in the Chalk. They are fairly common but cannot usually be identified. We collected a single specimen, which consisted of a mass of small bones dispersed through several cubic centimetres of Chalk. The bones are light brown and have shiny surfaces. They therefore contrast with the chalk and are easy to spot. Their colour distinguishes them from the invertebrate fossils that occur in the Chalk.

When we had finished collecting in the Upper Chalk we walked back towards the chalets. A series of clay and sand blocks begins about 50 m from the chalk cliffs. These blocks are up to 20 m high and are separated by deep channels. If you stand back to look at the blocks you will see that the beds of rock run vertically. The sediments were obviously deposited horizontally (p. 22), but since their formation the whole sequence has been tipped on to its side. The oldest rocks are at the southern end near the Chalk cliffs, and the youngest rocks are near the chalets. After the red clays of the Reading Beds, which are nearest the Chalk and contain no fossils, there is a ridge formed by the lower beds of the London Clay. At the base of this formation is a brownish-green sandy clay with small reddish clay balls and pebbles. A thin seam full of worm tubes (*Ditrupa*) occurs just above the base.

After collecting pieces of clay and worm tubes we continued walking towards the chalets. We went between the first and second pair of chalets and climbed 4 m up the cliff path. We then walked back about 10 m until we found a block of clay that is lighter than the surrounding rock. This block is easy to see, as it is very rich in fossils and its surface is usually kept clean by visiting geologists. The block is part of the Bracklesham Beds and is Eocene in age (about 40 million years old). It is full of shells, including *Turrilites sp.* and large forams – *Nummulites prestwichianus*. The molluscs are soft and difficult to collect, but the nummulites are harder. Cut out a block of this clay and take it home to sieve. The nummulites, which are up to a centimetre in diameter, can also be picked up from the soft sand at the base of the block.

Returning to the beach we walked to the valley between the two cafés. In the middle of this valley there is a high ridge with tussocks of grass intersected by footpaths over its crest. This ridge is made of grey clay that contains small whitish mollusc shells and very small nummulites. These are less than 2 mm across and are usually dark grey or black. After locating these deposits we collected a sample for sieving at home.

We then continued along the beach for about 200 m to an area littered with large white limestone boulders. These contain freshwater snails – *Limnaea* and *Planorbis*. The limestone of the cliffs dips towards the beach so that at the northern end of the bay it is very low, and above it there is a green clay

Commoner fossils from the Bracklesham Beds at Whitecliff Bay

First area collected:

Forams:	*Nummulites prestwichianus*
Molluscs:	*Turrilites*

Second area collected:

Forams:		*Nummulites variolarius, Alveolina*
Corals:		*Turbinolia*
Molluscs:	gastropods:	*Turritella, Clavilithes, Conomitra*
	scaphopods:	*Dentalium*
	bivalves:	*Corbula*

---

about 1 m thick that is capped with 5 cm of shelly sand. This contains well-preserved oysters (*Ostrea*) and a lot of shell fragments. Sharks' teeth have been found in this sand, but they are very rare.

Colwell Bay   Grid ref. SZ. 327879

From Whitecliff Bay we crossed the island via Newport and Yarmouth. We followed the A3054 from Yarmouth to Norton Green and then followed the signposted road to Colwell Bay, where there is free parking. Colwell Bay is a holiday resort, and there are cafés, shops and chalets near the beach.

Buses 11, 12 and 42 run here from Yarmouth. Ask for Brambles Chine and then walk down to the beach.

On reaching the beach we turned right, following the concrete pathway and then continuing along the shingle for about 50 m. Here the low cliffs contain a grey shell bed – the Oyster Bed – that is packed with oysters, clams and some gastropods. These are Eocene in age and are part of the Middle Headon

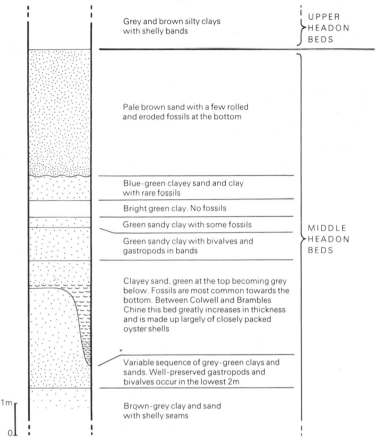

Grey and brown silty clays
with shelly bands

UPPER
HEADON
BEDS

Pale brown sand with a few rolled
and eroded fossils at the bottom

Blue-green clayey sand and clay
with rare fossils

Bright green clay. No fossils

Green sandy clay with some fossils

Green sandy clay with bivalves and
gastropods in bands

MIDDLE
HEADON
BEDS

Clayey sand, green at the top becoming grey
below. Fossils are most common towards the
bottom. Between Colwell and Brambles
Chine this bed greatly increases in thickness
and is made up largely of closely packed
oyster shells

Variable sequence of grey-green clays and
sands. Well-preserved gastropods and
bivalves occur in the lowest 2m

Brown-grey clay and sand
with shelly seams

1m

0

A vertical section of the Middle Headon Beds, Colwell Bay

Beds. After collecting in this spot for a short period we continued along the beach for another 50 m. This brought us almost to the next slipway. Here the Oyster Bed is much thinner, about 6 m above the beach. At beach level, however, the cliffs consist of dark grey sandy clays and are known as the Venus Bed. This bed contains the bivalve molluscs *Corbula* and *Sinodia*, which both look like clams. It also contains *Ostrea* and gastropods. Collect bags of this clay for sieving at home; in this way the soft shells will be protected. You may also find bits of crab among the sieved residues.

Commoner fossils from Colwell Bay

OYSTER BED

Molluscs: bivalves: *Ostrea*
          gastropods: *Urosalpinx, Nucula*

VENUS BED

Molluscs: bivalves: *Sinodia, Ostrea, Corbula*
          gastropods: *Viviparus, Pollia, Bonellitia, Borsonia, Tritonidea, Ancilla*

Alum Bay   Grid ref. S Z. 305855

After only an hour's collecting at Colwell Bay we returned to the A3054, turned right, and continued to Totland. From here we took the B3322 to Alum Bay. It is possible to park beside the road at Alum Bay but we parked near the top of the chair-lift and took the steps down to the beach. Arriving at the beach we turned left and walked to the end of the bay. We did not collect from the Chalk cliffs as we would merely have duplicated our earlier collection from Whitecliff Bay; but if you are only visiting Alum Bay, an hour or so of collecting from the Chalk cliffs should yield a good sample of the Upper Chalk fauna. The beds at this end of the bay are vertical, and the sand shows a sequence of different colours. Next to the Chalk there are reddish clays (Reading Beds), overlain by brown clays. The latter form the London Clay (Lower Eocene) and contain fossil molluscs that have a lustre like mother of pearl. We collected specimens of *Pinna*, which is a large bivalve mollusc, over 10 cm long, with a patterning of fine ridges crossed by weak growth lines. We also found

Alum Bay

Pholadomya, which is only 5 cm long and has stronger ridges and growth lines. We collected specimens of *Ostrea* and the clam-like *Panopea* as well as fragments of the long pointed gastropod *Turritella*. *Pinna* and the other molluscs occur in bands of hard rock that run through the clay.

We next collected from the greenish clay that outcrops at either side of the lower end of the chair-lift. This clay is part of the Barton Beds and is packed with shells, some sea urchin spines, scaphopods and occasional sharks' teeth. Although these beds are very rich, most of the shells are small, whereas at Barton on the coast of southern Hampshire opposite Alum Bay, where the same beds occur, there are many large molluscs that can be collected very easily.

We walked along the beach to the east of the steps for about 200 m and then climbed the gently sloping cliffs. These cliffs are very badly slipped, so that rocks from different levels are mixed together. Yellow and white limestones are known as the How Ledge Limestone and the Upper Headon Limestone;

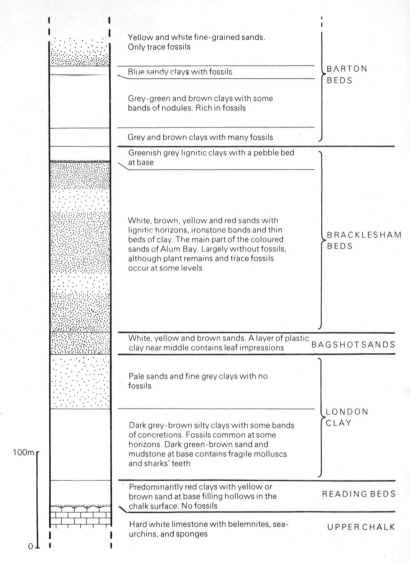

Yellow and white fine-grained sands. Only trace fossils

Blue sandy clays with fossils

Grey-green and brown clays with some bands of nodules. Rich in fossils

Grey and brown clays with many fossils

BARTON BEDS

Greenish grey lignitic clays with a pebble bed at base

White, brown, yellow and red sands with lignitic horizons, ironstone bands and thin beds of clay. The main part of the coloured sands of Alum Bay. Largely without fossils, although plant remains and trace fossils occur at some levels

BRACKLESHAM BEDS

White, yellow and brown sands. A layer of plastic clay near middle contains leaf impressions

BAGSHOT SANDS

Pale sands and fine grey clays with no fossils

Dark grey-brown silty clays with some bands of concretions. Fossils common at some horizons. Dark green-brown sand and mudstone at base contains fragile molluscs and sharks' teeth

LONDON CLAY

100m

Predominantly red clays with yellow or brown sand at base filling hollows in the chalk surface. No fossils

READING BEDS

Hard white limestone with belemnites, sea-urchins, and sponges

UPPER CHALK

0

A vertical section of the Eocene succession, Alum Bay

these are packed with freshwater gastropods. This part of Alum Bay is called Headon Hill, and many of the beds that outcrop on this hill bear its name. The grey clays are from the Lower Headon Beds and they contain masses of small shells. The cliffs are dangerous in wet weather but they are usually safe after a dry spell. We worked our way up the cliffs checking the limestone and the clays and collecting a few of the gastropods (p. 49). About halfway up the cliff there is an outcrop of How Ledge Limestone that is in position; in this limestone there is a band of greenish clay about 0·5 m thick. The top of this band is black with lignite, and the clay grows lighter in colour downwards. We spent half an hour looking through this blacker clay and found several mammal teeth. These are extremely small, as they are from mice, shrew-like mammals or opossums. However, the teeth and bones are jet black and glossy, so they contrast with the surrounding clay. We collected 3 kg of this black clay for sieving at home.

When we got this clay home it was dried in a slow oven for several hours. We then soaked it for half an hour and sieved it, using a kitchen sieve. The residue was again dried and sieved, and the resulting concentrate was dried and checked for mammal teeth. Twenty-two teeth were extracted from this small sample, and the bones of small reptiles (lizards) and fishes were also found.

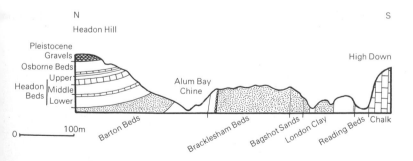

A cliff section of Alum Bay and Headon Hill

Above the How Ledge Limestone there are the Middle Headon Beds, which consist of dark brown and grey clays and sands. These include seams packed with fossil molluscs. Above this is the Headon Hill Limestone, which contains irregular bands of black lignite; these bands have yielded fossil mammals, but they are very rare. The Headon Hill Limestone contains freshwater molluscs.

Alum Bay is a very rich area for fossils. The Bay and Headon Hill join each other, but strictly speaking they should be regarded as two sites. In good weather the Bay is very pleasant and is well worth a full day's visit. This will allow time to concentrate on each site and very good collections can then be made.

Commoner fossils from Alum Bay and Headon Hill

UPPER CHALK

As Whitecliff Bay (p. 110)

LONDON CLAY

Molluscs: bivalves: *Pholadomya, Pinna, Ostrea, Panopea*
          gastropods: *Turritella*

BARTON BEDS

Molluscs: bivalves: *Corbula, Crassatella*
          gastropods: *Athleta, Sycostoma, Rimella*

LOWER HEADON BEDS (HOW LEDGE LIMESTONE)

Molluscs: gastropods: *Planorbis, Lymnaea*
Mammals

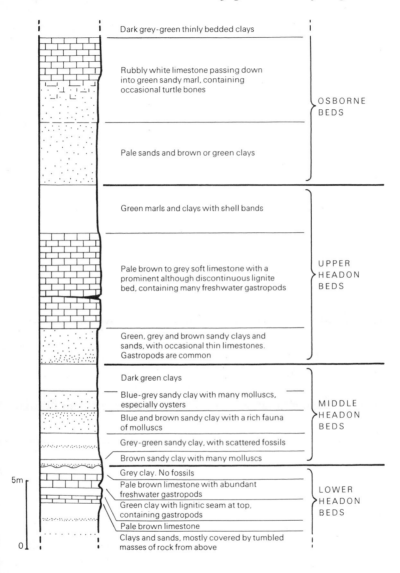

A vertical section of the Upper Eocene sequence on Headon Hill

MIDDLE HEADON BEDS

Molluscs: bivalves: *Trinacria, Ostrea, Corbicula, Sinodia*
             gastropods: *Melanopsis, Potamides, Batillaria,*
                         Ampullina

UPPER HEADON BEDS

Molluscs: gastropods: *Planorbis, Lymnaea*

Atherfield   Grid ref. S Z. 446798

From Alum Bay we returned to the B3322 and after 800 metres turned right towards Freshwater Bay. After Freshwater we turned on to the A3055 and continued to Shepherd's Chine (*not* Shippard's Chine), where we parked near the public footpath just after the Atherfield Holiday Camp. We followed the footpath through Shepherd's Chine to the beach. At the bottom of the steps we turned left and crossed a small stream. The cliffs are grey, with clearly defined layers, and there are many bivalve molluscs and ostracods (p. 54). The ostracods are very small, never more than 2 mm long, but they are present in vast numbers, so that they form thin layers in the grey rock. Each

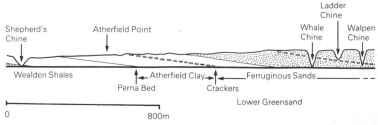

The Lower Cretaceous section between Shepherd's Chine and Walpen Chine

Upper *Gryphaea* Beds: red-brown sands crossing lower part of Walpen Chine

Walpen and Ladder Sands: greyish sands with a prominent bed of olive-green sandstone at base. Nodules contain ammonites and brachiopods. Sands contain oysters, brachiopods and wood

Upper Crioceras Beds: reddish-brown sands with prominent large concretions cutting across bottom of Ladder Chine. Nodules contain casts and moulds of bivalves and ammonites

Walpen Sands and Clay: dark sandy clays forming a prominent ledge around Whale Chine. Ferruginous nodules contain casts and moulds of bivalves and ammonites

Lower Crioceras and Scaphites Beds: reddish-brown sands whose top forms the bottom of Whale Chine. Large oysters and occasional echinoids; large ammonites in the harder bands

Lower *Gryphaea* Beds: brown sandstones with a band of about 5m above base with large oysters, and 'nests' of brachiopods

Upper Lobster Beds: pale grey-brown silts and sandy clays with crustaceans, bivalves, gastropods and ammonites

Crackers: very hard grey-brown concretions in brown clay. The concretions contain fossils

Atherfield Clay: grey and grey-brown silty clays. In upper 6m abundant crustaceans around small nodules associated with molluscs

Perna Bed: hard grey-green sandstone with many molluscs and corals

Black and grey laminated shales with occasional thin limestones. Thick yellow sandstone occurs at foot of Shepherd's Chine. Occasional bivalves and gastropods in shales; abundant bivalves in limestones

FERRUGINOUS SANDS

ATHERFIELD CLAY

WEALDEN SHALES

LOWER GREENSAND

WEALDEN

10m

0

A vertical section of the Lower Cretaceous rocks between Shepherd's Chine and Walpen Chine

ostracod looks like a tiny bivalve mollusc, and together they give the layers of rock a whitish, pitted appearance. Pieces of wood and leaves are also present in these rocks. They look like charcoal flecks and usually they cannot be collected, as they rub off the rock surface. A fern – *Weichselia* – is fairly common, and well-preserved specimens may occasionally be found. There are also trace fossils in these rocks.

Having collected some of these specimens we continued along the beach. We quickly began to find pyrite nodules and large oysters in the beach pebbles. We also found pieces of limestone derived from the Wealden Shales (Lower Cretaceous; about 120 million years old) which contained oysters and other bivalves. Near Atherfield Point we found the beach littered with large greyish green boulders from what is known as the Perna Beds. These contain oysters – *Exogyra* – that are up to 12 cm across and corals –*Holocystis* (p. 38) – which are spherical, with many star-like pits visible on their exposed surfaces. These boulders also contain worm tubes and the bivalve molluscs *Pholadomya* and *Pterotrigonia*. The beach pebbles containing beds of small bivalves and the pyrite nodules also became more abundant towards the point, and at the point we located the Perna Bed in position where it forms a ledge at the top of the beach. It extends across the beach and out to sea.

The landscape and distribution of rocks at Atherfield Point changes every year and even after every storm, so this description will only be valid for a short time; but you should have no difficulty in identifying the Perna Bed from the large oysters and corals that it contains.

Continuing past the point we again found large nodules containing bivalves, and about 50 m further on we found a huge mud flow that extended below the high-tide mark. This contains a mixture of fossils from the Atherfield Clay and from a bed called the 'Crackers' which contains large, extremely hard nodules. These occur higher up the cliffs, but fossils are more easily and safely collected at beach level. We found a good specimen of *Gervillella*, which is a large elongate bivalve,

A lobster, Meyeria, *embedded in the rock at Atherfield*

and the remains of a small lobster *Meyeria*. On our visit the collecting was unusually rich; in thirty minutes we found nine good specimens and sixteen other fragments of this small lobster. These clays also yielded several ammonite fragments.

After collecting from the Atherfield Clays we continued along the beach for about 800 m. Here the cliffs are very high, and along the base we found blocks of reddish-brown ferruginous sandstone. These are about 110 million years old, and they contain nests of brachiopods (p. 42) that are visible because they are whitish and therefore contrast with the sandstone. *Sellithyris* is a medium-sized brachiopod up to 2·5 cm long. It has a large foramen on the beak of its pedicle valve, and the back edge has two strong folds with a deep depression or *sulcus* between them. *Sulcirhynchia* is a smaller brachiopod with a wide shell carrying many fine ridges. To collect these brachiopods, take blocks of the sandstone with nests of brachiopods visible. Dry the blocks at home and wash them in fresh water. After this simple treatment the sandstone becomes soft and falls away, leaving the complete brachiopods or isolated valves. The sandstone also contains small specimens of the oyster *Exogyra*. There are many other fossils present, but these three genera are by far the most abundant. After extracting them the brachiopods should be soaked in fresh water for two or three weeks, as they contain a lot of salt that must be removed.

Along the cliffs we also found two fragments of a very large ammonite, but there was no way of finding out where these specimens had fallen from.

The walk back from Atherfield is very long and it is as well to be selective about the fossils that you collect. At high tide the point is cut off by the sea, so be careful, and try to time your visit so that you arrive with plenty of time to spare before and after low tide. Atherfield is a very rich area for collecting fossils and it has many different faunas; but it is hard to get to the site, so it is an excellent place to spend a whole day collecting. There are, however, no cafés or other facilities, so you will need a packed meal and drink.

Commoner fossils from Atherfield

WEALDEN SHALES

Molluscs:	bivalves:	*Filosina, Ostrea, Unio*
	gastropods:	*Paraglauconia, Strombiformis*
Plants:	ferns:	*Weichselia*

PERNA BEDS

Corals:		*Holocystis*
Brachiopods:		*Sulcirhynchia, Sellithyris*
Molluscs:	bivalves:	*Exogyra, Mulletia, Isognomon, Panopea, Gervillella, Gervillaria, Prohinnites, Sphaera, Protocardia, Astarte, Venilicardia, Noramya, Yaadia, Pholadomya, Pterotrigonia*
	gastropods:	*Fossarus, Globularia*
	ammonites:	uncommon but including *Prodeshayesites*

ATHERFIELD CLAYS

Fossils are rare except in the upper 6 m. The commonest are:

Molluscs: bivalves: *Panopea, Thetironia, Gervillella*
              ammonites: *Prodeshayesites, Deshayesites*
Arthropods: crustaceans: *Meyeria*

CRACKERS

Many fossils but very hard to extract.

Molluscs: bivalves: *Gervillella, Panopea, Pinna,*
                           Thetironia

FERRUGINOUS SANDSTONES

Fossils are abundant at several levels, but casual collecting usually yields only:

Molluscs: bivalves: *Exogyra*
Brachiopods: *Sellithyris, Sulcirhynchia*

Blackgang   Grid ref. SZ.491767

Leaving Atherfield we followed the A3055, and 800 m after Blackgang we parked in the hilltop car park on the right-hand side of the road. We then walked to the road cutting at the top of the hill. This cutting is in the Lower Chalk and here we collected ammonite fragments, a brachiopod and the oysters *Inoceramus* and *Liostrea*. This completed our day's collecting and we returned to Fishbourne to the ferry.

DAY 4   THE BRISTOL AND GLOUCESTER REGION

The day was fine and warm, with rain in the evening. We started from Bristol and worked in a loop to finish near Stroud. Collecting was from Mesozoic and Palaeozoic rocks and after a slow start we made a very large collection.

Aust   Grid ref. ST.565898

Take a heavy hammer and chisels.

We travelled west along the M4 and turned left just before the service area on the south side of the Severn Bridge, taking the B4461 and following signs to Aust Beach. We parked near the old ferry ticket office and then walked along the beach pathway towards the suspension bridge. We saw fossils in the

Cliffs at Aust, looking towards the suspension bridge

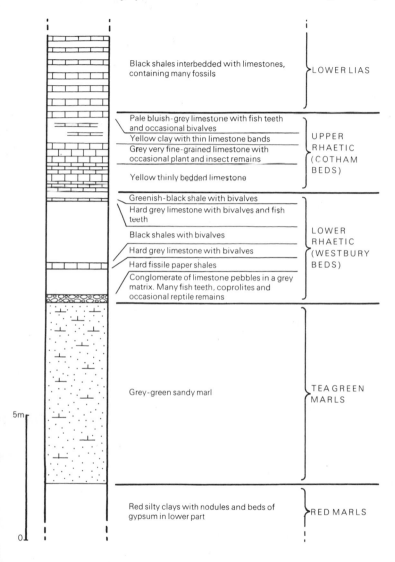

Black shales interbedded with limestones, containing many fossils — LOWER LIAS

Pale bluish-grey limestone with fish teeth and occasional bivalves
Yellow clay with thin limestone bands
Grey very fine-grained limestone with occasional plant and insect remains
Yellow thinly bedded limestone
— UPPER RHAETIC (COTHAM BEDS)

Greenish-black shale with bivalves
Hard grey limestone with bivalves and fish teeth
Black shales with bivalves
Hard grey limestone with bivalves
Hard fissile paper shales
Conglomerate of limestone pebbles in a grey matrix. Many fish teeth, coprolites and occasional reptile remains
— LOWER RHAETIC (WESTBURY BEDS)

Grey-green sandy marl — TEA GREEN MARLS

Red silty clays with nodules and beds of gypsum in lower part — RED MARLS

5m

0

A vertical section of the Upper Triassic succession at Aust Cliffs

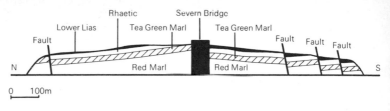

A cliff section at Aust

rocks and boulders all around the base of the bridge support. Aust is particularly important for the pieces of Aust Bone Bed which fall from high on the cliffs and litter the beach in early spring. However, this is a classic and very famous locality which is visited by school and university parties throughout the summer. As a result pieces of bone bed become more and more difficult to find, and during our visit we found only half a dozen small pieces. Aust should ideally be visited on a low or falling tide. It can be very cold here, as the winds funnel up the river, so wear heavy clothes.

The bone bed is grey or greenish in colour, but its surface may be yellowed by iron salts. The surface and broken faces show many pieces of bone and pebbles embedded in the rock. The bone is usually black and glossy. Also included are coprolites (p. 15) and many teeth. The bones belong to plesiosaurs and ichthyosaurs (p. 32), the fish *Birgeria* and the lungfish *Ceratodus*. There are many small teeth that are pointed at each end: these are from the shark *Acrodus* (p. 30). The conical pointed teeth usually have ridged surfaces; small ones are from *Birgeria*, larger ones from plesiosaurs and ichthyosaurs. Lungfish teeth are up to 4 cm long and are like triangular flattened plates, with large serrations along one edge and straight ridging of the tooth surface. From the blocks of bone bed that we collected we found examples of each kind of tooth, as well as bones and coprolites.

On the beach, and particularly north of the bridge, we also found slabs of grey limestone containing bones, teeth,

Aust Bone Bed, showing a fragment of Acrodus *fin-spine and a coprolite*

coprolites and pyrites. This is not true bone bed, but some of the best *Acrodus* teeth and spines (p. 30) may be found in these rocks, as well as the occasional larger bones. Although the Aust Beach is famous because of the bone bed, we also found blocks of Lower Liassic (about 190 million years old) rock containing many oysters as well as large blocks of Carboniferous Limestone with good corals (*Lithostrotion*).

Commoner fossils from Aust

AUST BONE BED

Molluscs:	bivalves:	*Mytilus, Schizodus*
Vertebrates:	fish:	*Ceratodus, Nemacanthus, Acrodus, Gyrolepis, Hybodus, Birgeria*
	reptiles:	*Ichthyosaurus, Plesiosaurus*
Trace fossils:		coprolites

PECTEN BEDS

Molluscs:	bivalves:	*Chlamys, Protocardia, Schizodus, Rhaetavicula, Pleurophorus*
Vertebrates: fish:		*Acrodus, Birgeria, Gyrolepis*
Trace fossils:		coprolites

THIN LIMESTONES

Molluscs:	bivalves:	*Pseudomontis, Modiolus, Pleurophorus*
Vertebrates: fish:		*Gyrolepis, Birgeria*

Notes on the fallen blocks

1. Pale green sandy blocks that do not contain fossils come from the 'Tea Green Marl' which is common along the whole section.

2. An irregular conglomerate of rounded Carboniferous Limestone (grey-black), Tea Green Marl (greenish-grey), coprolites (black or brown), and bones (black or brown) in a green matrix is the Aust Bone Bed.

3. Dark green limestone that is very rich in fossils and has a peculiar 'churned up' surface (bioturbation) and rusty patches due to the decomposition of pyrites is the Lower Pecten Bed. The Upper Pecten Bed is similar but is paler and lacks the rusty stains.

4. Pale buff limestones with bedding planes closely covered with the bivalve mollusc *Pleurophorus*. Thin limestones occurring within black shales above and below the Lower Pecten Bed.

5. Hard cream-coloured calcareous mudstone – the Insect Bed. This contains rare insect remains, *Eustheria* (an ostracod-like crustacean), and remains of the moss *Naiadites.*

6. Blue-grey calcareous mudstone containing thin darker

flakes arranged chaotically within it – Crazy or False Cotham Marble.

7. Pale buff limestone with bedding planes often covered with fossils – basal Lower Lias fossils include *Lima*, *Modiolus*, *Pleuromya*, *Atreta*, *Protocardium*, *Liostrea*, *Oxytoma*.

8. Soft white or pink gypsum.

Lydney   Grid ref. SO.653018

Make sure you have wellington boots.

Leaving Aust we returned to the M4 and crossed the suspension bridge into South Wales. We left the motorway on the A466 towards Chepstow and followed this road through Chepstow before changing on to the A48 to Lydney. On entering Lydney village we turned right, taking the B4231, which is signposted to the 'Station and Industrial Estate'. We followed this road for about 1½ kilometres until, with the canal on our right, we reached a gate across the road. After parking we walked down to the beach, where we turned left (northeastwards), crossing the slipway and walking carefully across the mud and grass. This part is very variable. It may be dry, but during our visit it was very muddy, so that wellington boots or good hiking boots are essential. This site must be visited at low tide. After about 400 m we reached an area where the base of the cliffs was clear of vegetation, and here we found blocks of coarse red sandstone scattered under the low cliffs. Some of these blocks contain a lot of mica and many darker clay pebbles. These rocks show marked bedding and may already be splitting along the bedding planes. This sandstone is Devonian in age. It contains the spines and scales of armoured fishes (p. 29). The fish remains are chalky white when broken, but their unbroken surfaces may be white or bluish. Some spines are several centimetres long and the scales are up to 2 cm across. Scales may have pitted surfaces, or they may have a fine spongy texture. Good specimens may be found on

Fallen blocks on the beach at Lydney

weathered surfaces, or they can be exposed by splitting the rocks. Good specimens should be photographed at the site, as they are fragile and often break up before they can be carried home. In less than an hour we collected eight specimens of spines and scales.

Cinderford   Grid ref. SO.644155

Returning to Lydney, we turned right on to the A48 and continued on this road as far as Blakeney, where we turned left on to the Cinderford road (B4227). We drove through Cinderford, and about $1\frac{1}{2}$ kilometres past the town we visited an open-cast coalmine on the left of the road. Here we collected fossil plants in the shales and clays that are found on the tip heaps. The black glossy plant remains contrast with the grey shale and are easy to see when exposed by splitting the shale along the bedding planes. Bits of fossil wood, leaves and roots were collected (p. 59). At the time of our visit this coalmine was closing, and the open-cast pit was being filled and land-

Fossil leaves on shale from Cinderford

scaped. It is therefore unlikely that collecting will continue to be possible at this site. However, the Forest of Dean contains many small coal-pits, and you should be able to get good plant fossils from the tip heaps of any of these.

Fossil plant remains from Cinderford include: *Neuropteris, Alethopteris, Pecopteris, Mariopteris, Annularia, Cardaites.*

Plump Hill   Grid ref. SO.661171

We continued along the A4151 for about a kilometre, then turned right on to the A4136. After $1\frac{1}{2}$ kilometres we passed two lakes on the right and an old iron-mine on the left. After this, on the outskirts of Plump Hill, there are two quarries on the left-hand side of the road. We visited the second one, which is just after Quarry House. This quarry contains Carboniferous Limestone and Lower Limestone Shales. The weathered surface of the shale is yellowish, and fossils stand out from the surface. These fossils include crinoid stems and corals.

At this point our day's collecting had been only moderately successful. We had collected only a few specimens from each site and we did not have a great variety of fossils. However, we were now on the way to a very rich Silurian locality.

Longhope   Grid ref. SO.695193

We continued along the A4136 through Longhope, turned left on to the Ross-on-Wye road and after $1\frac{1}{2}$ kilometres turned right on to the A40 towards Gloucester. After another $1\frac{1}{2}$ kilometres we turned right and then after only 1 kilometre we turned on to an unsurfaced track, where we parked. We walked to the end of this track, where there is a stretch of quarries extending for over a kilometre.

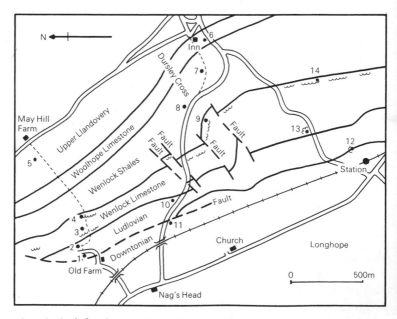

A geological sketch map of the Silurian rocks of the Longhope area, showing the position of the main collecting sites (see p. 138)

These quarries are in the Wenlock Limestone, which is
Middle Silurian in age and is therefore about 425 million years
old. When collecting in Wenlock Limestone it is best to ignore
the rock face and to collect from the floor of the quarry. The
weathered limestone surface shows a great variety of fossils,
including trilobites, solitary and colonial rugose and tabulate
corals (p. 39), including *Halysites*, bryozoans, brachiopods and molluscs. We picked through the rubble, selecting
the best blocks. The surface of many blocks was coated with a
soft yellowish rock, but during preparation this was easily
removed with a small pick after soaking the blocks in water. So
if the exposed fossils are good, the block may be worth collecting even if there is a covering of this soft yellow rock.

There are many quarries in this area and a lot of them are
worth visiting. In quarries that are still being worked you must

Halysites *from Longhope*

ask the site manager before approaching the workings, but disused quarries are often better for collecting, as the rubble will have been weathering for longer.

Commoner fossils from Longhope
Corals: *Favosites, Heliolites, Halysites*
Brachiopods: *Atrypa, Leptaena, Resserella*
Trilobites: *Dalmanites*

Some other quarries in this area (see the map on p. 136)

1. The left-hand bank of the minor road that runs eastwards and then northwards from Old Farm, about 100 m beyond the farmhouse (Grid ref. SO.68442072). This is in the Upper Ludlow (Upper Silurian) Series and contains many brachiopods.

2. A sunken track through the Middle Ludlow Series at Grid ref. SO.68482076. This again contains many brachiopods.

3. A line of disused quarries north-east of Old Farm (Grid ref. SO.68552075). These are in the Wenlock Limestone and contain corals and brachiopods.

4. A disused quarry north-east of Old Farm (Grid ref. SO.68662072). This is in the Wenlock Limestone and contains corals and brachiopods.

5. A cutting near the abandoned May Hill Farm (Grid ref. SO.69022109). This is in the Upper Llandovery beds and contains brachiopods and corals.

6. A road cutting on the A40 south of Dursley Cross (Grid ref. SO.69841997). This is in the Woolhope Limestone and contains brachiopods.

7. Exposures along the line of the old Roman Road south-west of Dursley Cross (Grid ref. SO.69621997). This is in the Wenlock Shales and contains brachiopods.

8. A roadside stream culvert west of Dursley Cross (Grid ref. SO.69302009). This is in the Wenlock Shales and contains brachiopods.

9. Quarries in Sculchurch Wood (Grid ref. SO.69191993). These are in the Wenlock Limestone and contain corals and brachiopods.

10. A roadside culvert south-east of the Nag's Head Inn (Grid ref. SO.68832011). This is in the Lower Elton Beds resting on Wenlock Limestone and contains brachiopods and trilobites.

11. A road cutting on the south side of the A40 east of the Nag's Head Inn (Grid ref. SO.68702022). This is in the Leint-wardine Beds and the Whitcliffe Beds, which are faulted against Downtonian sandstones and shales. Ludlovian beds consist of flaggy siltstones with limestone bands and contain many brachiopods and a few trilobites.

12. A quarry behind Longhope railway station (Grid ref. SO.69101906). This is in the Whitcliffe Beds, overlain by Ludlow Bone Bed and Downton Castle Sandstone. Whitcliffe Beds contain abundant brachiopods.

13. A quarry beside a lane north-east of Longhope station (Grid ref. SO.69231932). This is in the Wenlock Limestone and contains abundant brachiopods.

14. A line of quarries running south from the lane north-east of Longhope station. This is the group of quarries we visited.

Gilbert's Grave road cutting north of Northleach
Grid ref. SP.146195

Leaving the quarries we turned eastwards along the A40 and went through Cheltenham, turning left on to the A429 at Northleach. We continued along this road for $5\frac{1}{2}$ kilometres and shortly after the signpost to Notgrove (stay on the A429) we passed through a road cutting on a hill top. We parked beside the road and walked back to this cutting.

Here the yellowish rocks are part of the Inferior Oolite, which is Middle Jurassic (175 million years) in age. This road cutting is very rich in fossils. Large pieces and some complete specimens of the sea urchin *Clypeus* are scattered around in the

Clypeus, *a sea urchin, from Gilbert's Grave*

rubble below the cutting, and other specimens can be dug out
of the soft rock in the upper part of the cutting. Brachiopods
are also abundant. Bivalve molluscs, gastropods and regular
echinoids were collected, but we got relatively few of these.

Commoner fossils from Gilbert's Grave

Brachiopods:		*Stiphrothyris, Rhactorhynchia*
Molluscs	bivalves:	*Trigonia, Liostrea, Homomya,*
		Pseudolimea, Myophorella,
		Pleuromya, Pholadomya
	gastropods:	*Chomatoseris*
Echinoids:		*Clypeus*

Crickley Hill   Grid ref. SO.928163

Returning on the road towards Cirencester, we turned right on to the A40 at Northleach and left on to the A436 towards Gloucester, later changing to the A417 Gloucester road. We turned right off this road towards Cold Slad and parked when the road became unsurfaced. We walked up this track to Crickley Hill.

Fossils are found near the top of Crickley Hill in the soft rocks under the harder blocks of Lower Freestone. There is also a scatter of rocks all down the slope, and good weathered specimens occur in the scatter.

The fossil-bearing rock at Crickley Hill is called the Pea Grit, because it contains small disc-shaped elements about the size of a pea. These are called pisoliths. Their origin is not certain, but each one has a nucleus formed from algae. Fossils are very abundant in the rocks of Crickley Hill, and we spent over an hour collecting along the vertical rock face near the top of the hill. Specimens from this site are easy to clean at home using small picks and needles.

Brachiopods are particularly common and range in size from *Pseudoglossothyris*, which is up to 3 cm long, to the small *Crania*, which is less than 1 cm long. We also collected sea urchins – *Holectypus* (p. 56) – complete tests and spines of regular sea urchins, bivalve molluscs, fragments of crustaceans, crinoid stems, branching bryozoans and corals. We finished collecting for the day at six o'clock.

Commoner fossils from Crickley Hill

Plants:	algae:	*Girvanella*
Sponges:		*Diaplectia, Lymnorella* (4 species)
Corals:		*Adelastraea, Anabacia, Chorisastraea, Cosmoseris, Cyathophyllia, Dimorpharaea, Donacosmilia, Goniocora, Isastraea, Latimeandra, Microsalena, Montlivaltia, Oroseris, Placophyllia, Phylloseris, Stephanocaenia, Thecoseris, Thecosmilia, Thamnasteria*
Annelids:		*Serpula*
Bryozoans:		*Actinopora, Apsendesia, Berenicea, Ceriocava, Diastopora, Haploaecia, Heteropora, Kololophos, Multiclausia, Proboscina, Stomatopora, Theonoa*
Brachiopods:		*Aulacothyris, Crania, Pseudoglossothyris, Rhynchonella* (4 species), *Terebratula* (3 species), *Plectoidothyris, Zeilleria*
Molluscs:	bivalves:	*Avicula, Cardium, Corbis, Eopecten, Hinnites, Lima, Lopha, Modiolus, Pecten, Pholadomya, Placunopsis, Plicatula, Trichites, Trigonia*
	gastropods:	*Cerithium, Crossotoma, Emarginula, Natica, Nerinea, Nerita, Neritopsis, Pileolus, Pleurotomaria, Puncturella, Rimula, Rissonia, Solarium*
	belemnites	
	nautiloids:	*Cenoceras*
Echinoderms:		
	crinoids:	*Pentacrinus*

echinoids: *Acrosalenia* (2 species), *Cidaris* (3 species), *Diplocidaris, Galeropygus, Hemipedina* (5 species), *Pedina, Pseudodiadema, Polycyphus, Pygaster, Plesiechinus, Stomechinus*

starfish: *Goniaster*

Arthropods: crustaceans: *Eryma*

DAY 5 THE YORKSHIRE COAST

The north and east coast of Yorkshire is an excellent area for collecting fossils and it marks the farthest north that we collected. The Yorkshire moors, which are very high, fall away along the coast as steep cliffs. These coastal sites are difficult to collect from, as the fossil-bearing rocks are usually visible only at low tide, and in several places, including Port Mulgrave and Whitby, there is a real danger of being cut off by the tide unless you are very careful. The weather during our trip was very good. We arrived in the evening and spent the night at Port Mulgrave. On the following day we worked southwards, visiting sites as we went. We used the low-tide period at Port Mulgrave and then collected from three sites that are not dependent on the tide before attempting another coastal site.

Port Mulgrave   Grid ref. NZ. 798177

Take a sledge hammer and chisels to this site.

We followed the signposted road off the A174, continuing through the village of Port Mulgrave and following the cliff road to its end at the northern arm of the bay. This site can only be collected from shortly before and after low tide. The cliffs are 130 metres high and very steep. There are no facilities on the beach, so if you plan to spend several hours collecting be sure to take food and drink with you.

We took the cliff path to the beach and turned left, walking in a north-westerly direction for about a kilometre. Here the beach is made of grey shale flats, and we saw many ammonites, belemnites and *Inoceramus* in this shale. Fossils are very difficult to collect from the shale but there are many nodules embedded in it. These are lighter grey, and some of them are stained pink. The nodules contain ammonites, bivalve molluscs and small gastropods that may be pyritized (in which case they will be golden in colour). Large crystals of pyrites also occur in bands in many of the nodules. The nodules vary in size from 10 cm to 1 m across. We found ammonites exposed on the surface of some nodules, but we got our best specimens by cracking open the smaller nodules. Most of the nodules contain fossils, but they are all extremely hard, and a sledge hammer is needed to crack most of them. Alternatively use the blunt end of your hammer. Do not use the chisel or pointed end, as the hard rock will break off splinters of steel and this can be dangerous. If we were sure that a nodule contained a fossil we wrapped it and took it home for careful preparation. The best ammonites were found in small nodules and were exposed as a serrated or shiny band around the nodule. We found very few of these nodules (three), but early in the spring the smaller ones are more common.

Different genera of ammonites occur in different parts of the shore. *Dactylioceras* is found furthest from Port Mulgrave harbour and *Harpoceras* occurs nearest the port. This is because the shales dip gently, so that beds of different ages occur in different places. Continuing past the point – Thorndale Shaft (Grid ref. NZ.798181) – we entered Brackenberry Wyke. Here the shore is made of huge flats of reddish ironstone. Fossils were common but difficult to collect, so we contented ourselves with photographing the richest areas. There were many large 'scallops' and abundant trace fossils in the ironstone. Here we had to watch the tide very carefully, as this bay can be cut off at two points by the rising water. After looking over this area we returned to the port and crossed to the beach on the south-eastern part of the bay. Here we again found

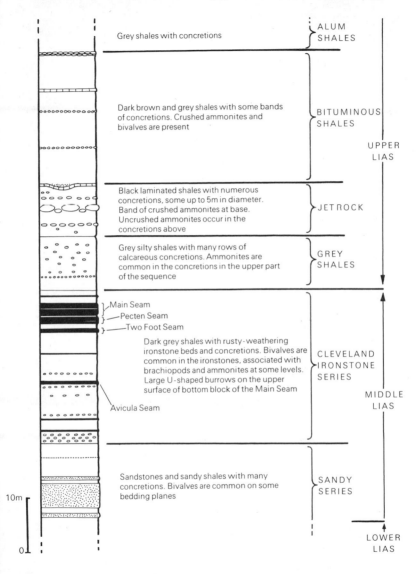

A vertical section of the Upper, Middle and Lower Lias sequence exposed on the foreshore between Staithes and Port Mulgrave

Brackenberry Wyke, Port Mulgrave, showing ironstone beds on the foreshore

shales with nodules containing ammonites. We collected here for a short period and spent a few minutes on the small stretch of shingle near the harbour, where some of the pebbles had ammonites either enclosed or exposed on their surfaces. This shingle can be checked while the tide is rising, as there is safe access to the cliff path.

Commoner fossils from Port Mulgrave

IRONSTONES IN BRACKENBERRY WYKE

Brachiopods:		*Tetrarhynchia*
Molluscs:	bivalves:	*Pseudopecten, Pleuromya, Pecten, Gryphaea, Pholadomya*
	ammonites:	*Pleuroceras*
Trace fossils:		*Rhizocorallium*

GREY SHALES

Molluscs: ammonites: *Dactylioceras*
               belemnites: *Belemnites*

JET ROCK (on the west side of Port Mulgrave and also amongst the nodule-bearing shales to the east)

Molluscs: ammonites: *Harpoceras, Hildoceras, Eleganticeras*

Notes on the Fossils

Rhizocorallium is a very abundant trace fossil that forms horizontal U-shaped tubes.

Port Mulgrave is well known for its ammonites. *Dactylioceras* has a strongly ribbed shell. The ribs are fine and continue around the venter. The whorl cross-section is almost circular.

Hildoceras is characterized by the strong ridge and grooves that encircle the venter. Its shell carries a few low curved ribs.

Harpoceras has a more flattened shell, and its venter carries a strong ridge. The first whorl of *Harpoceras* is large and partially covers the smaller inner whorls. The ribs of *Harpoceras* are fine and strongly flexed in the middle of the face of each whorl. Crushed specimens of *Harpoceras* are abundant in the shales.

The belemnites retain their shape in the shales, and we also found a specimen of *Belemnoteuthis* which is a squid-like cephalopod. Its black glossy remains were flattened in the shale, forming a triangular area with a slightly rippled surface.

Port Mulgrave and many of the other Yorkshire sites are visited regularly by students, school parties and amateur collectors throughout the summer. As a result the number of ammonites exposed decreases rapidly, and the area is far less

Belemnoteuthis *flattened in the shale at Port Mulgrave*

rich than it was only a few years ago. However, we visited the area late in the season and were still able to find good specimens, so it is probable that each collector will find a few ammonites in the area.

Dactylioceras, *an ammonite, exposed by cracking a nodule*

Hackness Quarry   Grid ref. SE.96559065

From Port Mulgrave we travelled south-east along the A174 and in Whitby we turned south on to the A171 Scarborough road. In Scalby we turned right following signposts to Hackness. Just after Hackness village we turned right again, following the road to 'Low Dales and High Dales'. Only 250 m after this turning we visited a fenced-off quarry on the left-hand side of the road. This quarry contains sandstones of Middle Jurassic age, which are about 155 million years old. Fossils are found in the top two metres of this sandstone, and we found pieces of this near the bottom of the quarry face. We were also able to reach the upper bed by climbing among the bushes on the right (west) side of the quarry face. This sandstone is permanently wet and muddy and as a result it does not appear at first glance to contain any fossils. There are, however, nests of brachiopods and many other fossils occurring in concentrations within it. We broke open the sandstone blocks and looked over the broken surfaces for fossils. Having

Hackness Quarry

established in this way that we were collecting from the correct beds, we took blocks of the sandstone away for careful checking and preparation of the contained fossils.

Commoner fossils from Hackness Quarry

Brachiopods: *Rhynchonella, 'Terebratula'*
Molluscs: bivalves: *Gryphaea, Chlamys, Pleuromya*
            ammonites: *Kosmoceras, Quenstedtoceras,*
                      Peltoceras, Pseudocadoceras,
                      Longaeviceras, Grossovaria,
                      Hecticoceras
          belemnites

Betton Farm Quarry   Grid ref. TA.002856

We returned to the main road, turned right and then followed this road towards Forge Valley and to East Ayton. In East Ayton we turned left into Castle Lane, then right into Moor Lane and left on to Racecourse Road (B1262). After about a kilometre we passed Betton Farm on the right, and shortly

The face at Betton Farm Quarry

after this we parked on the left-hand side of the road at the gates of a disused quarry.

In this quarry the lower part of the face contains few fossils, but the upper part consists of a fossilized coral reef and therefore contains many fossil corals. These are usually massive, and bands of coral can be seen on broken faces of the rock. We also found large gastropods (*Bourguetia*), bivalves and solitary corals. Brachiopods are present but rare. On the floor of the quarry there were piles of rock which included many weathered pieces. In these rocks we found five large gastropods and a good specimen of *Lopha*.

Commoner fossils from Betton Farm Quarry

Corals:		*Isastraea, Thamnasteria,*
		Thecosmilia, Montlivaltia
Brachiopods:		' *Terebratula* '
Molluscs:	bivalves:	*Lima, Chlamys, Lithophaga, Lopha,*
		Navicula
	gastropods:	*Littorina, Bourguetia,*
		Pseudomelania, Trochotoma,
		Nerinea, Natica
Echinoderms:	echinoids:	*Hemicidaris, Paracidaris*

Scarborough Castle Hill   Grid ref. T A. 047815

We then continued along the B1262 to Scarborough, where we took the coast road around the base of Castle Hill. We parked on the cobbled section of the road. A low wall runs beside the road around the base of the cliffs, and behind this wall there are piles of large boulders that have fallen from the cliffs. These cliffs are very dangerous, so avoid obviously dangerous parts and observe the warning signs. The boulders contain *Pinna* and beds showing trace fossils indicating extensive burrowing. *Pinna* looks like a large mussel and it may sometimes be seen quite clearly, as its shell is dark and therefore contrasts

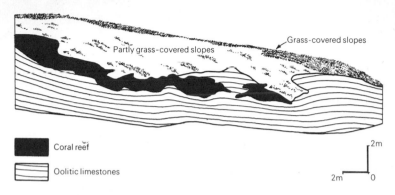

Coral reef

Oolitic limestones

2m

2m   0

A sketch of the section at Betton Farm Quarry showing the relationships between the reef masses and the surrounding limestone

Bourguetia, *a gastropod, from Betton Farm Quarry*

with the yellow sandstone. Specimens of *Pinna* can be chiselled out of these blocks.

Grisethorpe   Grid ref. T A. 085842

Leaving Scarborough on the A165 we continued to Grisethorpe, where we turned left, following signs towards Dennis's Camp. We parked on the cliff top and took the cliff path to the beach. On the beach we found grey nodules containing masses of *Ostrea*. Plant remains occur in yellow sandstone on the beach, and we found that they become increasingly common towards the headland at the north-west end of the bay – Yons Nab. They are black and form seams in the sandstone. Leaves and fruit may be visible and we found one large block in which plant remains were clearly exposed. Because of the tide we were unable to reach the Grisethorpe Plant Bed, which occurs just around the point from Grisethorpe Bay.

Grisethorpe Bay is particularly interesting, as large sections of the cliff are made from Boulder Clay. During the Pleistocene much of the Northern Hemisphere was covered by thick ice sheets which moved southwards and retreated northwards periodically. These ice sheets consisted mainly of glaciers that moved over the land and sea beds; they plucked rocks and rock particles from the surface as they passed. When the glaciers finally melted this load was deposited as Boulder Clay. The glaciers had transported their loads over great distances, so that rocks and boulders can now be found in Boulder Clay many hundreds of kilometres from the areas where these rocks normally occur. On the beach at Grisethorpe there are many rocks and boulders representing kinds of rocks that only occur a long way from their present site. Of course these boulders are not fossils. They are, however, very good evidence of the most recent major geological event to occur before the start of written history and are therefore of interest to the collector.

5   *Working on your Collection*

PREPARING SPECIMENS

Once you get your specimens home the second phase of making a collection begins. This is the preparation of the specimens, and it involves removing the fossils from the surrounding rock. The methods and tools that should be used depend on the nature of the surrounding rock. It will take time for you to develop all the necessary skills, but with practice excellent results can be obtained with very little equipment.

When the surrounding rock is soft it can be removed by gentle brushing or washing. When the matrix is harder, more vigorous methods are needed. Usually you can trim the rock away slowly using a small hammer and narrow chisels. Careful use of these will get to within a few millimetres of the fossil. Obviously at this stage it is important to know the likely shape and orientation of the specimen so that you do not damage it. Sometimes a well aimed blow of the hammer will 'jump' the fossil clean of the rock, but more usually a thin veneer of rock will remain attached to the specimen. This can be removed using mounted dissecting needles, a fine knife or any other suitable instrument with a sharp point. Do not scrape the rock away, because this will damage the surface of the fossil. Instead, press the end of the tool at right angles to the surface, and then small fragments should break away cleanly from the surface of the fossil. Occasionally you will find fossils that cannot be cleanly separated from their surrounding rock. This results from the method of preservation, and there is nothing you can do about it; so be satisfied with what you have got.

The cleaning process can be greatly speeded up by using an engraving tool known as a Burgess Vibrotool. This costs

The results of sieving clay from Headon Hill, Isle of Wight

several pounds, so until you are certain that you will continue collecting it is not worth the expense.

Chemical solution can be used to clean some fossils. For example, if the fossil is a bone and is preserved in limestone, the limestone can be removed by using a weak solution of acetic acid. In some limestones you may find fossils that have been silicified. This difference between the matrix and fossil allows the use of dilute hydrochloric acid (car battery acid is suitable). Remember that acids are dangerous and should be handled with care. Always try to wear rubber gloves and take care not to get acid on your clothes, skin or in your eyes.

If you want to extract small fossils from a large sample of unconsolidated rock you can use a sieving technique (see also p. 119). First dry the sample slowly in a warm dry atmosphere or in a slow (Gas No. 1) oven. When the sample is completely dry soak it in water for a time so that it breaks down. You can then wash the sample through a suitable sieve, using a gentle spray of water. Professional geologists use brass wire sieves of various mesh sizes, but these are expensive and in most cases much cheaper equipment works just as well. A very cheap sieve can be made from a pair of tights or an old nylon stocking supported on chicken mesh. Alternatively you can make a sieve using zinc gauze fastened to a wooden frame. You should always wash the sieve thoroughly between each sample to avoid contaminating (mixing) your next sample. You may sometimes find samples that will not break down sufficiently to pass through the sieve. This may be because the sample was not completely dry to start with, in which case you should repeat the drying process. There are, however, some samples that cannot be broken down simply by soaking them. There are several ways of dealing with this, using various chemicals, but the simplest method is to boil the dried clay in water with a small amount of a liquid detergent for about half an hour to an hour. This can be a messy operation.

If you break a specimen during collection or during preparation you will need a suitable glue to repair it. A quick-setting, non-flexible glue such as Uhu is suitable for small specimens, but for heavier specimens a quick-setting epoxy glue such as Araldite Rapid is necessary.

Some specimens require special treatment to preserve them. Fossils collected from sea cliffs invariably contain salt water. After a time the water will evaporate, depositing salt, which can cause the specimens to break up. It is therefore important to soak all such material in fresh water for a week or so to remove the salt before drying and storing. Fragile or powdery specimens need strengthening; and this can be done using Uhu or Durofix dissolved in acetone or ethyl alcohol.

The most difficult problem that you will have to deal with is pyrites disease. Some specimens made of pyrites are stable, but others will decompose to a mass of powder in a few years. There is no way of permanently preventing this type of decay but various techniques will slow the process down. For example, the specimen can be immersed in ammonia solution, which will get rid of all the oxidation products of pyritization. These products turn red in ammonia. Then coat the specimen carefully with cellulose solution and allow it to dry. This will exclude air and damp and will greatly prolong the life of the specimen.

MAKING A COLLECTION

Storing a large collection of fossils can be a problem. It will certainly take up a lot of space. However, you can build storage units out of cardboard boxes, tomato trays or plastic containers; and if you are good at carpentry you can make your own storage cabinets as you need them.

At first you will probably gather an assorted collection of fossils with all groups represented, but as you gain experience and skill you may wish to collect only fossils of a particular animal group, those from a particular rock formation or those from a small area. Whatever the kind of specialization you choose, the important thing is to label each specimen with at least the locality and rock formation from which it was collected. This data, giving as much detail as possible, considerably increases the value of the collection. You may, for example, have collected from a temporary exposure that was not seen by anyone else, or you may find a completely new species, but in either case the scientific value of the specimen will be greatly reduced if there is no information about the site where it was collected.

Finally, if you think you have discovered a new fossil you should take it to a museum so that it can be examined by an expert. Many new or rare fossils are discovered every year by amateur collectors, so there is always a chance that you can do the same.

BOOKS AND OTHER SITES TO VISIT

British Caenozoic Fossils
British Mesozoic Fossils
British Palaeozoic Fossils
All available from the British Museum (Natural History),

Cromwell Road, London, SW7. These three excellent hand-books include stratigraphical data and figures of most of the common fossils occurring in Britain. They are cheap and will be very useful for preliminary identification of fossils.

Invertebrate Fossils, R. C. Moore, C. G. Lalicker and A. G. Fischer (McGraw-Hill, 1952)
This contains detailed descriptions of the groups of fossil invertebrates. Terminology is explained and hundreds of invertebrate species (mainly from North America) are illustrated and described.

A Directory of British Fossiliferous Localities (Palaeontographical Society, 1954)
This gives a county-by-county listing of fossil sites but does not provide very precise directions or grid references. Many localities are omitted and the directory is rather out-of-date.

Geologists' Association Guides
These can be obtained from *The Scientific Anglian*, 30/30a St Benedict's Street, Norwich, NOR 24J. They provide itineraries of particular areas and are inexpensive.

Handbooks of Regional Geology of Great Britain
Obtainable from the Museum of Practical Geology, Exhibition Road, London, SW7, and from Her Majesty's Stationery Office bookshops. These guides are available for most areas of Britain and they include details of fossils, geological sections, maps and descriptions of some sites.

Index

The numbers in *italic* refer to illustrations